D0916009

59774

681.113 Smi
Smith, Eric, 1940-.
Striking and chiming clocks

DATE DUE			

MONTROSE LIBRARY DISTRICT
MONTROSE, CO   81401

MONTROSE LIBRARY DISTRICT   MA   STACKS
681.113 SMI   Smith, Eric,
Striking and chiming clocks :

1 11 0000437133

STRIKING AND CHIMING CLOCKS

STRIKING AND CHIMING CLOCKS

Their Working and Repair

Eric Smith

David & Charles
Newton Abbot   London

Arco Publishing, Inc.
New York

59784

Line drawings by Eadan Art

British Library Cataloguing in Publication Data

Smith, Eric, 1940–
  Striking and chiming clocks: their working
  and repair.
  1. Clocks and watches – Repairing and
  adjusting
  I. Title
  681.1'13   TS547

  ISBN 0-7153-8662-X

© Text and illustrations: Eric Smith 1985

All rights reserved. No part of this
publication may be reproduced, stored
in a retrieval system, or transmitted
in any form or by any means, electronic,
mechanical, photocopying, recording or
otherwise, without the prior permission
of the publishers.

Published in the UK by
David & Charles (Publishers) Limited
Brunel House, Newton Abbot, Devon

Published in the USA by Arco Publishing, Inc.
215 Park Avenue South, New York, NY 10003

Library of Congress Cataloging in Publication Data

Smith, Eric, 1940–
  Striking and chiming clocks.
  Bibliography: p.
  Includes index.
  1.Clocks and watches – Repairing and adjusting.
  2. Chiming clocks.   I. Title.
  TS547.S593   1985   681.1'13   84-21749

  ISBN 0-668-06422-6

Phototypeset by Typesetters (Birmingham) Limited
Smethwick, Warley, West Midlands
and printed in Great Britain
by Butler & Tanner, Frome and London

CONTENTS

INTRODUCTION

Were we to enter into a detail of all the nick-nacks which have been introduced into the striking part of a clock, we might write a whole quarto volume on the subject.

Abraham Rees, *Clocks, Watches and Chronometers*, 1819–20

If that was the position in the early 1800s, developments since Rees would make the undertaking even more formidable today – and perhaps, as he half implies, unnecessary. But there is a good, practical reason for the present book. Over many years of repairing, restoration, and indeed designing and making of clocks, I have received more questions and carried out more work on what may be called the 'sounding' side than on the 'going' side. Yet I have found relatively little explanation and illustration in print, and I imagine that others with similar interests have found the same. The appeal of the 'sounding' side is wide. It embraces all the many attractions of clock-work, hard as these are to define, together with, at least for me, musical intrigue. A clock that does not 'sound' its best, whether in accuracy or in tone, is to me as defective as one that limps along 'out of beat'. The chime wrongly phased, or the hammers damping the gongs, are as offensive as the dial with hands of the wrong size, or the poor beast that has to be wound up every three days because it needs a new spring.

There are several reasons for the prevalence of problems on the sounding side. In the 1920s and 1930s, the heyday of the mass-produced chiming clock in Britain, new clocks presumably gave no undue trouble. Certain standard designs evolved, with self-correcting work, which were virtually foolproof and are still followed in smaller production lines today. Eventually, however, this generation of clocks required attention and then often they would be switched to 'silent' and the expense of repair deferred, or left unwound, or in some other way wisely or foolishly condemned to silence. Now these clocks are being inherited by a generation that seems to have a liking for sounding rather than silent clocks – as indeed there were always those who liked to have a 'tick' added to the noiseless synchronous machine.

There is even a fully electronic chiming clock, providing a choice of tunes, on the market. The inherited and rediscovered clocks – including many, of course, from earlier centuries – are presented to the amateur repairer in increasing numbers, for the local clock-shops who once would have seen to the matter have largely gone out of business and the multiples will not touch them.

Besides the 'bulge' of clocks from the 1920s and 1930s, the range of striking and chiming arrangements from other periods is enormous, as Rees observed. They start from when striking (at least) was appropriate and functional to the house clock, since most rooms did not have a clock at all. Even if it might be 'silenced' or 'repeated' or, indeed, functioned only on demand at night, the sounding was once almost as important as the hands (or single hand) to indicate the passing of time; whereas for much of the last two centuries it has been increasingly ornamental. What proportion of house clocks, even without the pre-war glut, were made to strike or chime, one cannot begin to estimate. Undoubtedly it is large. The longcase clock, time-teller for the household at large, is virtually never without a strike, save in the area of select regulators, which are not really house clocks. To our forefathers, striking was usually essential. Chiming – Chapter 1 attempts to distinguish striking from chiming – was not; quarter-chiming during normal running (as opposed to on the pull of a cord) was relatively unusual in the seventeenth and eighteenth centuries, but always available as a luxury. It spread in the 'second' or luxury clock of the time, the bracket clock, rather than in the longcase; but by the late eighteenth century it was a popular option. Nonetheless, there can be no doubt that in Britain the near coincidence of the Great Exhibition of 1851 (where gongs made a notable appearance in chiming clocks) and the installation of Big Ben in 1859 greatly strengthened the popularity of chiming, which was to become almost standard (as were its tunes) in the first half of the twentieth century.

In saying that, despite the great volume and variety of sounding devices in domestic clocks and despite the fact that this side of the clock is at least as prone to give trouble as the going side, there exists no manual for the repairer, particularly the novice, to learn about the mechanisms, their working and maintenance. I am not thinking here so much of the sumptuous antique – restorers have recently received much help in that area from Harvey and Allix's *Hobson's Choice*. I am thinking more of reasonably standard middle-price and cheaper bracket, mantel and longcase clocks of the past 250 years or so; in other words, clocks which are the bread and butter of amateur and professional repairers and which range widely from individual design and craftsmanship to mass-production and from modest collectors' pieces to standard domestic articles. They exist in astonishing variety and are often set up with apparent ignorance or even indifference as to how they should run and sound. Moreover, a mechanical fault on the sounding side very often stops and may even damage the going side.

I have attempted to summarise the principal systems, mention some common variations and offer some notes on repair and setting up, but always concentrating on the main road. Apart from the impossibility of including all variations – it is most unlikely that anyone can even be aware of all of them – certain arrangements are deliberately excluded from the book because of their variety, the value in some instances of the relatively few clocks where they are found, and because they are discussed, if with little practical detail, elsewhere. Examples of such systems are Dutch and Roman striking, ship's striking, Surrerwerk, and all old English grande sonnerie and repeating arrangements. Some of the books mentioned in Further Reading will be of help here. Guesswork and improvisation are folly in the restoration of valuable antiques; I can lay down no line, but would suggest erring on the side of caution and seeking specialist advice where there is doubt.

Many repairs and techniques apply equally to both striking and going sides – bushing and pivoting, repairing and replacing wheels, attending to mainsprings; these are some of the commonest jobs. It is assumed that the reader is aware of what is needed in these areas and has some knowledge of how to do it. In fact, workshop skills and techniques do not bulk very large here at all – the reason for this may be illustrated by the following anecdote of some years ago when I had to look at a Northern longcase clock belonging to a friend. It had a day-of-the-month disc below the hour hand and this was plainly turned every twelve hours by a pin on the hour wheel. It also had a 'moon' disc in the break-arch, but how this was turned I could not imagine – there was no pinned wheel anywhere near, no lever and no stud to indicate that one was missing. There was, however, an inexplicable threaded hole in the false-plate to the dial. I eventually discovered that this hole was the site for a stud bearing a simple pivoted lever with a weighted pawl, the fork of the lever being raised by the pin on the hour wheel, causing the pawl to advance the moon disc; and I made something up which hopefully was reasonably typical of what I later learned was called a 'deer's foot trip', from its shape, though I have never seen this clearly drawn or photographed. Thus the problem is far more often to decide what is missing and has to be made, than just how to make it. The story also shows that with an old clock you must sometimes, having done your research, be content to make a part which functions and appears in character, as well as being decently made – riveted rather than soft-soldered, for instance – and at least you are not then participating in any attempt at deception.

Because of the importance of this problem of missing parts, the photographs have been chosen to give a reasonably comprehensive survey of types and to show their principles and working, rather than to illustrate exceptions to rules which may not themselves be too well known. I am greatly indebted, as always, to Ann Allnut, for the hours spent in impossible positions trying to capture on celluloid with one lens what in reality required stereoscopic vision; also to Derek Roberts in a similar capacity. I am grateful besides to the many who have allowed their clocks to be exposed for this purpose, whether specially or in the course of being repaired. I hope that my own sketches pin-point certain details which may not be clear in a photograph; they do not pretend to do more.

1

ELEMENTS OF SOUNDING MECHANISMS

Striking, Chiming or Musical?

To many people, 'striking' and 'chiming' are vague and virtually inter-changeable terms, though they probably know what they mean by 'musical' clocks and are not far wrong in this direction. The confusion is not new. In the eighteenth century, 'chiming' was regularly used for what we now call 'musical', and 'strike' has always been used for 'chime' – so much is clear from clock dials. As there is some variation even in horological circles, it is as well to define how these terms will be used in this book.

I shall use 'striking' to be indication, usually but not always on one bell or gong, of the hour or of the quarters by a second train where there are but two trains. Normally, one blow equals one hour, though there are complications. Except in a few instances where there is strik-ing from the going train (principally the passing strike of one blow at the hour caused by the falling of a hammer tail from a pin or cam on the minute-hand arbor), the striking train is thus the second train in a striking clock and has its own power supply. Usually, but not always (and in fact seldom in thirty-hour longcase and lantern clocks with an endless rope or chain and having trains side by side), the striking train is on the left when seen from the front. This layout took some time to become established (see, for instance, Plates 6–7) in clocks of longer durations. I would still call a clock 'striking' even if there were two or more bells or gongs, and even if the quarters were indicated by this same train. Such clocks – mostly ting-tang on two bells – seem best called 'quarter-striking'. Some two-bell clocks strike only the hours – a European Continental fashion which appeared in the late nineteenth century – and these are familiarly known as 'bim-bams', one bim-bam being equivalent to one blow. The cuckoo clock with two piped notes and a blow on a gong is similar in principle. One ting-tang, on the other hand, normally equals one quarter, and in a ting-tang quarter striker the hours are usually sounded on only the lower bell; in quarter chiming, however, with a third train for quarters, four ting-tangs are usually sounded at the fourth quarter before blows on a different bell for the hour.

I shall take 'chiming' to involve the use of a third train (usually on the right) and more than one bell or gong, sounding at least every quarter, with or without the hour as well. (Quarter striking or chiming, with the hour, is known as 'grande sonnerie'.) The whole chime is what is played at the hour, plus the last part of what is played at the third quarter. This is because a chime is made up of a number of sequences – as we shall call them for simplicity's sake, for the art of change-ringing to which clock chimes are partly related has a terminology of its own which is not strictly appropriate – and one sequence represents a quarter of an hour. During the whole hour, if you add all the quarters together on this basis, ten sequences are required to sound, and these are in practice normally five sequences repeated; the sixth and first sequences are the same, occurring at a quarter past the hour and again at the end of a quarter to the hour. The length of chiming is thus proportionately related to the periods represented: half past is two sequences, whilst an hour is four sequences, and so on. The five sequences can only be played through in their set order, which presents certain problems should it be required to 'repeat' a quarter at will. In practice, in the age of repeating (broadly speaking, from the end of the seventeenth to the beginning of the nineteenth century), quarters were very often indicated by one sequence only (such as the ting-tang) in simplest form, because of the difficulties posed by a more elaborate tune.

Musical clocks normally involve a further train of gears or, if not 'further', a train reserved for the music itself. And they play a complete tune at intervals, usually every three or four hours, though many combinations exist. The length of this tune has no significance in telling the time, although there may be a different tune for each day of the week. The musical side is thus more decorative than an indication of time passed; it is a large and specialist subject, related mainly to expensive and relatively rare pieces.

Basic Principle and Counting Systems

In the simplest terms, conventional striking and chiming mechanisms are motors – trains of gears held constantly under power and let off by a lifting piece which links them to the going train. In certain cases, they can be let off manually by depressing a lever at will.

A slow-running wheel early in the train has, round its circumference, pins or cams on which rest the tails of a pivoted hammer or hammers. When this wheel revolves, the hammer tail drops off the pin and the hammer head at the other end of the shank strikes the bell or gong.

The speed is regulated by friction (and thus by how much power is applied to the wheel train), by the weight of the hammer and strength of its return spring, and by a fan or fly which is the last member of the train and whose effect is dependent on its mass and surface area. Generally a light fly of large surface area – an air-brake – is nowadays preferred, and it may have adjustable vanes so that the speed can be varied. Simple and much heavier flies were used in older clocks where reliance was more on mass as a brake, and these massive simple flies persisted well into the nineteenth century on traditional longcase movements. The sumptuous memorial or baronial longcase clock of 2.7 to 3m (9 to 10ft) in height became fashionable in the middle of that century and had a tendency to employ complex adjustable flies. Edmund Beckett Denison, afterwards Lord Grimthorpe, who was responsible for London's Big Ben, laid very great stress in his writings on what may be called the modern concept of the fly as a light air-brake, and was caustic on the older style. The change is of some importance as it corresponds to that of the strike from a simple functional indicator to a decorative and musical luxury.

All striking or chiming devices are required to perform an extending series of blows from one to twelve (possibly also with one at each half), or an extending series of sequences from one at the first quarter to, usually, four at the fourth quarter. Once the train has been released, it may run on remorselessly in a programmed order, regardless of any sudden change in the position of the hands; or, ideally, it will refer back to the position of the hands to determine how long its run shall be. In the twentieth century there is the compromise possibility that it will run like an automaton but only for a limited period, when it will be held up, at the third quarter, until the hands catch up and it is self-corrected at the hour. So far as I know, this device of self-correction has not been applied to hour striking, and of course it would be of little use there since it would allow the clock's striking and hands to disagree for up to eleven hours rather than a mere three-quarters of an hour.

Basically, there are only two counting systems and they can be used equally well for hours or quarters. They are the countwheel, which runs its own sweet way but may be self-corrected; and the rack which, if correctly set, can strike or chime only as indicated by the hands. We will consider the principles of each in turn.

The Countwheel
The countwheel system is also known as the locking-plate system because the notched disc which is in fact its programme appears to lock the train, although in actual fact it only determines when the train may be locked by other means. There are many forms of countwheel, but the commonest is a disc with raised or projecting sections or pins set

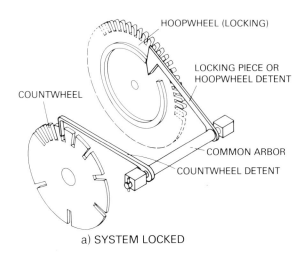

a) SYSTEM LOCKED

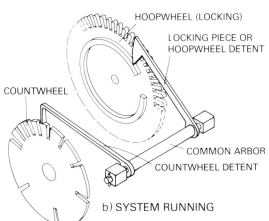

b) SYSTEM RUNNING

Fig 1   The countwheel principle

out proportionately to the number of blows required in twelve (occasionally twenty-four) hours, or of sequences in an hour. Some rare seventeenth- and early eighteenth-century countwheels incorporate both hours and quarters on one large disc; but these will not come the way of most repairers.

Thus a simple twelve-hour countwheel is based on 78 divisions (the number of blows struck in twelve hours), or on 90 divisions if one is also sounded at each half hour. A quarter-chiming countwheel is based on 10 divisions, there being ten chime sequences to the hour regardless of how many blows – in total, including spaces, always the same in each sequence – there may be to the sequence. When the train is released, a lever or detent rides over the projections and falls into the first notch; which coincides with a revolution of the locking wheel or

14

hoopwheel, where another detent stops the train. One revolution of this wheel is made for each blow or sequence (Fig 1). Once that section of countwheel has been passed it can be followed only by the next; there is no possibility of repetition. Ten must be followed by eleven even when the hands show nine; the only possible correction is to run the strike round to nine again or to move the hands on to eleven without allowing the clock to strike. The same applies if, for some reason – usually because the owner has advanced the hands without letting the clock chime at each quarter – half past is chimed instead of a quarter past. In the latter case, however, a self-correcting device may lock and silence the chime until the next hour.

The Rack

The great advantage of the rack system is that accuracy in relation to the hands is assured. This is due to the fact that its controller or counter, equivalent to the countwheel, is driven not by the sounding train but by the going train, being fixed to or linked to the pipe carrying the hour hand (hours) or to the minute wheel revolving once an

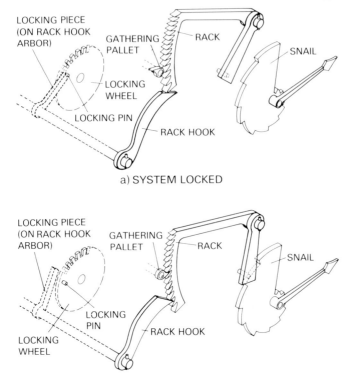

a) SYSTEM LOCKED

b) SYSTEM STARTED RUNNING

Fig 2   The rack principle (pin locking)

15

hour (quarters) (Fig 2). This device is known as the hour or quarter snail and is continuously curved or marked in steps in the shape of a whorl for the hours, or divided into four proportionate raised sections for the quarters. The rack itself is a toothed arm able to fall onto this snail and, as one tooth of the rack represents one blow or sequence, the distance of its fall governs how many blows are struck. Since the tail of the rack falls onto the stepped snail, positioned according to the hands, the striking must correspond to what is shown by the hands, and the same goes for the quarters. The rack is gathered up by a pallet on the end of the locking wheel's arbor (revolving, again, once per blow or sequence; though locking may be by other means in the rack system) until, when all its teeth are up, the train is locked. The rack is held up, and prevented from slipping whilst it is being gathered, by a rackhook which functions as a click or pawl to a ratchet, sliding over the sloped face of each tooth but catching on the upright face.

Combination Systems
The uses of these two counting systems are various and may best be seen as:

| Striking Only | Striking and Chiming | |
	Striking	Chiming
countwheel	countwheel	countwheel
rack	rack	rack
	rack	countwheel

Sequence of Events

Let-off and Warning
As has been said, let-off is most often produced by the raising of a lifting piece connected to the locking by the motion work. There may be one pin (hours), two pins (halves as well) or four pins on either the central cannon wheel (clockwise) or minute wheel (anti-clockwise) with which it engages (Fig 3). In older and country clocks – and indeed in many Black Forest and cuckoo clocks – where there is no central arbor or only one hand driven outside the plates by a pinion on the greatwheel, the let-off may be from a pinned or starwheel with twelve points, on which rides a lifting piece in the form of a long hook riding up the pins or teeth. But let-off from a wheel revolving once an hour is much commoner. The quarters are let off in a similar manner from four pins on the minute wheel, and here special arrangements have to be made for letting off the hour. Various alternative, mainly nineteenth-century, systems employ some form of flirt. Here the lifting pin, or a

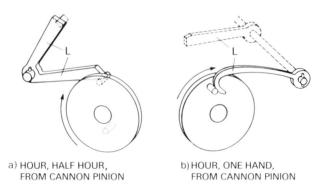

a) HOUR, HALF HOUR,
FROM CANNON PINION

b) HOUR, ONE HAND,
FROM CANNON PINION

L = LIFTING AND WARNING PIECES

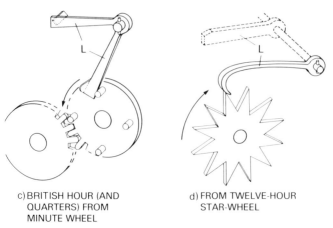

c) BRITISH HOUR (AND
QUARTERS) FROM
MINUTE WHEEL

d) FROM TWELVE-HOUR
STAR-WHEEL

Fig 3   Lifting arrangements

pivoted piece in its path, draws back a sprung lever which is released on the hour or quarter and flies up or across the clock, knocking out the locking piece (Fig 4).

Usually the activating motion wheel is on the front plate behind the dial, but during the nineteenth century it became fashionable, especially on the European Continent and in the USA, to place these wheels between the plates. The minute wheel – as in any number of the cheaper carriage clocks – was usually screwed to the back of the front plate with a stud, but some American clocks gave this wheel a full-length arbor and mounted it between the plates as if in the main train. This required modified let-off arrangements. The main advantage of these changes was not ease of adjustment – indeed, they can cause many practical difficulties – but the saving of space in the overall depth of the clock and the ability to mount cheaper dials without dial pillars.

17

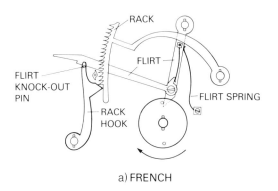

a) FRENCH

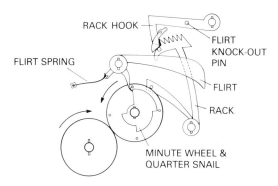

b) BRITISH QUARTERS

Fig 4   Flirt let-off arrangements

Let-off and warning are related parts of the process which starts the clock striking or chiming. Let-off, the actual release, is instantaneous – a lifting piece falls off the motion-wheel pin or cam which has been raising it, or a spring is released to knock out the locking device. But the whole process is not instantaneous, and it makes demands on the energy of the striking train. A train must be unlocked against its own power, a rack must fall correctly onto its snail, a countwheel detent must be raised from its notch in a wheel necessarily directly or very nearly directly under the pull of a spring or weight. In practice, therefore, with certain calculated exceptions, unlocking is a gradual and sliding process and is moved as far up the train as practicable, in view of the lesser friction met there. Furthermore, always for the countwheel system and usually, but not always, for the rack system, the trains are released a few minutes before sounding begins and held up by a subsidiary locking – the warning piece acting on the warning wheel – until the exact moment for let-off (Fig 5). Then the lifting piece (which, besides freeing the train, immediately raises the warning stop)

18

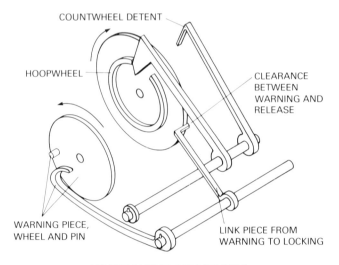

COUNTWHEEL DETENT

HOOPWHEEL

CLEARANCE
BETWEEN
WARNING AND
RELEASE

WARNING PIECE,
WHEEL AND PIN

LINK PIECE FROM
WARNING TO LOCKING

a) THIRTY-HOUR COUNTWHEEL

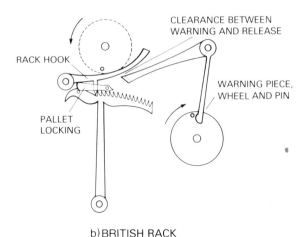

CLEARANCE BETWEEN
WARNING AND RELEASE

RACK HOOK

WARNING PIECE,
WHEEL AND PIN

PALLET
LOCKING

b) BRITISH RACK

Fig 5   Clearance at warning

and warning piece fall away together and the train runs freely.

Warning is in some arrangements a necessity, in some a safeguard, and in others customary rather than essential. Generally, however, if this provision were not incorporated, there would be the danger of relocking, perhaps dangerously on the countwheel itself which, as we have seen, does not lock at all before the train has really got under way, and the counting device could fail to operate properly. You may have had some experience of this with a repeating carriage clock. Correctly designed, this has no warning, the repetition lever stopping the fly while the button is depressed, so that the rack can fall fully; but, dressed up from a non-repeating model to enhance its value, it may not have this arrangement and may well have warning. In the latter case, with or without warning, the rack cannot fall reliably because the train may start to run and gather it prematurely. The stopping of the fly replaces warning in the true repeater because warning would make the strike inoperable for some minutes before the hour; however, this subtlety was commonly ignored in clocks other than carriage clocks and they may have authentic repetition cords and levers for the hour but also have warning and no means of stopping the fly. (In this simple sense, of course, any rack striking clock can be made a 'repeater'.) Warning is not essential in a rack clock, even a non-repeater, provided that the train is set up to have a long enough run before gathering starts, so that the rack will always fall fully. Warning was never used, as far as I know, in the curious and long-lasting Comtoise clocks, even those with a repeat facility. Neither was it usable in those, not necessarily repeating, clocks which employ a flirt let-off. Here the force taken from the going train and stored in the flirt's driving spring is sufficient to unlock the train fully and, if properly set up, the fall of the rack and counting are quite reliable and coincide precisely with the hour. Actually they coincide as far as convention accepts it to be necessary, for domestic clocks commonly begin chiming on the hour, so that their actual hour stroke may be almost half a minute late. Special arrangements have been made for important clocks with chiming to begin their *strike* on the hour, but for all *chimes* other than at the fourth quarter to be on time. Finally, I would stress that if warning is present, it must be set up correctly when the clock is assembled, or unreliable striking is almost sure to result (see page 167).

Locking

The three common arrangements (Fig 6) for arresting the sounding train when not performing are, first, a lever (locking piece) intercepting a pin on the locking wheel – in most eight-day clocks, the fourth wheel, whose arbor is extended to carry the gathering pallet in rack striking. Secondly, a lever drops into a slot in a cam or hoop, this slot, on older clocks, consisting of a gap in a hoop riveted to the side of the hoopwheel (the second wheel in thirty-hour and the third in eight-day clocks). In both these arrangements the locking wheel revolves once for each blow or sequence struck. Thirdly, there is the rack system, where various uses have been made of the gathering pallet itself to halt the train once the rack is fully gathered up, the classic British arrangement being for the pallet's tail to catch on a pin projecting from the front or back of the end of the gathered rack, leaving the train free to run (usually only to warning) once the rack is released at let-off. Whatever the system, locking is subject to critical adjustments. It must decisively halt the

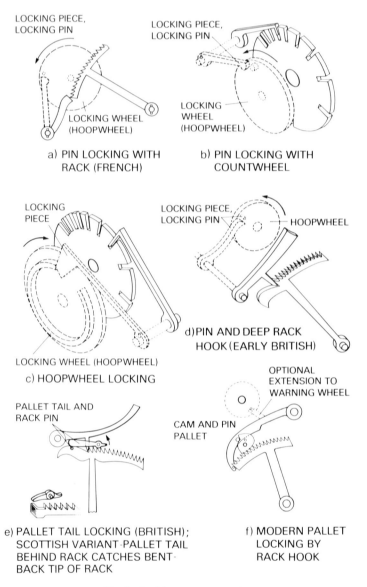

LOCKING PIECE,
LOCKING PIN

LOCKING WHEEL
(HOOPWHEEL)

a) PIN LOCKING WITH
RACK (FRENCH)

LOCKING PIECE,
LOCKING PIN

LOCKING
WHEEL
(HOOPWHEEL)

b) PIN LOCKING WITH
COUNTWHEEL

LOCKING
PIECE

LOCKING WHEEL (HOOPWHEEL)

c) HOOPWHEEL LOCKING

LOCKING PIECE,
LOCKING PIN

HOOPWHEEL

d) PIN AND DEEP RACK
HOOK (EARLY BRITISH)

PALLET TAIL AND
RACK PIN

OPTIONAL
EXTENSION TO
WARNING WHEEL

CAM AND PIN
PALLET

e) PALLET TAIL LOCKING (BRITISH);
SCOTTISH VARIANT-PALLET TAIL
BEHIND RACK CATCHES BENT-
BACK TIP OF RACK

f) MODERN PALLET
LOCKING BY
RACK HOOK

Fig 6   Locking arrangements

sounding train, yet it must not demand so much energy from the going
train as to hinder let-off when the time comes to sound.

The principle of the hoopwheel or locking cam was used on the oldest
striking systems and has continued until the present day, though now
it is used mainly on the chiming side. On all forms the locking piece
rides freely on the cam until it falls into the notch. In early British

21

5S794

striking the locking piece's face has both a straight edge to catch the edge of the hoop as the hoop notch approaches, and a curved edge to ease its removal from the notch at let-off; in forms from the last century on, it is more usual for the process to be reversed, the notch moving away from the locking piece which, on falling into the notch, catches and holds. As this occurs at a later stage in the chiming train, there is less problem in lifting the locking piece for let-off. This is the commoner form on early mass-produced German and American skeleton-plate movements and on modern chiming movements.

In locking with a pin on the wheel, again the locking piece may collide with the pin, as seems to be universal in French clocks, and in eighteenth-century British clocks, or, typically in modern chiming clocks, it may catch the pin as the latter comes up from behind. This form of locking is found in both rack and countwheel systems. In the countwheel system, of course, the locking piece falls into place when the countwheel detent meets a notch on the countwheel, whilst in rack devices the locking piece is on the same arbor as the rackhook, which makes a similar distinctive movement as it comes below or behind the fully gathered rack. Occasionally the pin and hoop methods are combined, so that when the locking piece falls into the hoopwheel notch a projection from it also catches the warning pin on the next wheel in the train. It was common in the early stages of pin locking, ie the first fifty years of the eighteenth century, for the first tooth of the rack to be preceded by a deep notch; when the rackhook encountered this, its fall correspondingly set the locking piece, on the same arbor, in the path of the locking pin. Grimthorpe, writing in the second part of the nineteenth century, notes as characteristically French – but wholly admirable – a system whereby the rackhook protrudes into the movement, like the warning piece, and operates on the warning wheel's pin, being made to fall by a similar notch in the rack. This system does not seem to be widespread, but clearly anticipates a modern liking for using the warning pin for locking as well as warning in rack systems (see also Fig 28).

The pallet system, no less than the others, existed in many forms and has continued in use until the present day. In its modern form, the gathering pallet comprises an elliptical cam and a projecting pin. The pin gathers whilst the cam, which has a straight-faced notch in it, engages with a pin or bent-up projection on the rackhook when the latter has fallen into place beneath the gathered rack. The purpose of the elliptical cam is to raise and lower the rackhook so that the teeth are properly held during gathering. Ratchet teeth are not usually employed and the hook does not slide like a pawl as in the traditional form where the pallet tail locks on the rack itself, although even in older clocks there are occasional instances where the pallet tail locks

22

onto the rackhook. A still clearer example is in the circular-rack arrangement adopted for some modern cuckoo clocks; here the gathering pallet is shaped into a hook to engage with a pin on the rackhook (Plate 48).

Hour Let-off

The question of hour let-off in chiming clocks has already been touched on in connection with the precise timing of the hour strike. From this it

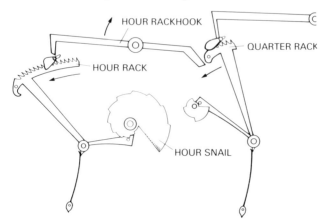

Fig 7 Hour let-off in quarter-rack chiming. As the quarter rack falls to its lowest at the fourth quarter, it knocks out the hour rackhook, and the hour rack falls, its tail striking the snail

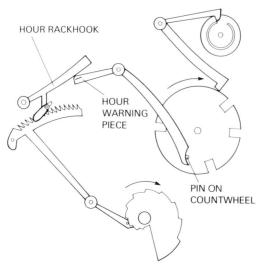

Fig 8 Hour let-off in eighteenth-century countwheel chiming. The pin on the count-wheel raises the hour warning piece which releases the hour rackhook so that the train runs to warning and, when the warning piece is released, striking begins

23

might have been inferred, correctly, that the normal (though not universal) arrangement is for the strike to be let off by the fourth quarter's chime. This has been done in various ways. In the rack mechanism for chiming it is general for the specially deep fall of the rack required at the fourth quarter – as dictated by the quarter snail – to dislodge the hour rackhook (Fig 7). The hour warning piece goes right across the clock and is held 'off' only when the quarter rack is fully gathered. The piece is sprung to give constant warning and is raised against this spring by a pin at the foot of the quarter rack; thus in a rack chimer the hour is at warning whenever the quarters are being struck, but of course the hour does not sound until the hour rack is released and its gathering pallet is freed and, further, until the quarter rack is once again fully gathered so that the hour warning is released. This arrangement involves a somewhat sensitive balancing of springs and so an alternative has been much advocated, but not very often adopted. This is for a pin (behind or in front of the lifting pins) on the quarter snail to work a pivoted lever which dislodges the hour rackhook at the fourth quarter.

Such a system is in fact similar to that used on older clocks with countwheel chiming, where there is a pin in the fourth quarter of the countwheel which works on a pivoted lever. This lever is, in effect, the usual hour lifting piece for a rack strike. It activates in turn the hour warning and the release of the hour rack by raising the rackhook

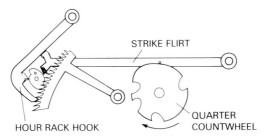

STRIKE FLIRT

HOUR RACK HOOK   QUARTER COUNTWHEEL

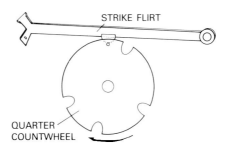

STRIKE FLIRT

QUARTER COUNTWHEEL

Fig 9   Hour let-offs in modern countwheel chiming

24

(Fig 8); the shape and length of the tip of the lever are critical, determining the break between chiming and striking. In the present century this pin is usually replaced by the use of an extra-high fourth-quarter section on the chime countwheel, this section alone being able to raise a lever, the strike flirt, high enough to set off the strike warning and release the rackhook (Fig 9). Sometimes a pin in the countwheel encounters a downward bulge in the strike flirt, with the same effect. Other systems used include a special pallet on the countwheel arbor, at the fourth quarter; this presses on an extended hour rackhook tail. There have also been successful applications of an independent hour lifting piece raised by the cannon pinion; these require careful setting up if the hour is to sound reliably at the right moment.

Other Features

Half Hours
The arrangement for striking a blow at each half hour also comes in many forms. This was a European fashion from at least the eighteenth century, but it did not become popular in Britain until the latter part of the nineteenth. Quite possibly its presence in all the carriage clocks and semi-standard round French movements in ornate cases which were exported to Britain at that time led to its introduction in British clocks; but the British seem always to have preferred hours alone, or with quarters, to the single half-hour blow. The main means of securing the blow are passing strike, where a hammer is simply lifted and let fall by the motion wheel independently of the hour train; widened notches in countwheels; or modifications to rack striking whereby the rack does not fall, or falls only one tooth, at the half hour. This latter arrangement, which can be achieved in several ways, was occasionally extended to provide the sounding of quarters also by the lower rack teeth – a form of quarter striking.

The countwheel set-up is virtually self-explanatory. The wheel, as has been said, is based on 90 divisions, the additional 12 (to the usual 78 for the hours) being spaces at which the train is released but the countwheel detent falls immediately so that the locking wheel can revolve only once and only one blow is sounded. Occasionally spaces for half-past twelve and half-past one are excluded, so that the confusion of three successive single blows at midday and midnight does not occur.

Where one rack tooth is released, the bottom rack tooth is short and the half-hour lifting pin in the motion work permits only this short tooth to be released. Some such system must be used if locking is by the pallet tail, since without release of the rack the train cannot run. Alternatively, and very commonly in French clocks, the lifting piece releases the rackhook but a lever worked from the minute motion

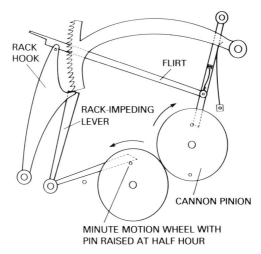

RACK HOOK

FLIRT

RACK-IMPEDING LEVER

CANNON PINION

MINUTE MOTION WHEEL WITH PIN RAISED AT HALF HOUR

Fig 10   French half-hour strike without rack fall

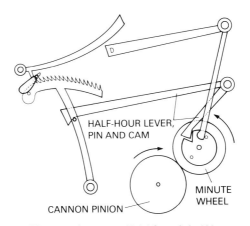

HALF-HOUR LEVER, PIN AND CAM

MINUTE WHEEL

CANNON PINION

Fig 11   Nineteenth-century British rack half-hour work

wheel is moved below the rack so that it cannot fall although the train will run until the rackhook falls back into place after one blow (Fig 10). In repeating carriage clocks with flirt release, the flirt has two notches which permit varying depth of engagement with the rackhook. The shallow one does not knock it out sufficiently to release the rack fully or at all, and so only one blow is struck. This is a fairly precarious arrangement and is again accompanied by the lever moving across below the rack to prevent it from falling at the half hours.

This half-hour blow came to be adopted in Britain on rack clocks although, curiously, Grimthorpe writing in 1874 said: 'I have never seen this in any English clock. Indeed, the English house clock-makers seem determined to lose every bit of trade rather than allow any single improvement to be made here, and so they are losing more and more yearly.' And, here, a special arrangement, to which nineteenth-century writers refer, was advocated (Fig 11). This involved the addition of a notched cam on the minute wheel, the notch corresponding to the half hour. On it rested a cranked lever whose other end was hooked; at the half hour the lever fell and the hook caught a pin in the rack, permitting it to fall only one tooth. As was generally agreed, this is a simple and infallible method; but it does not seem to have been widely used in practice and most British half-hour strikers use the less satisfactory short rack tooth method.

Half-hour striking is a frequent source of trouble, and one of the places where patient adjustment is often required in overhauling a striking clock.

Repeating
Repeating clocks are of two kinds: those where there is no normal running strike or chime but where the clock will sound when a cord is pulled which both winds and lets off the mechanism; and those which ordinarily strike and/or chime and can be made, by pulling a cord or pressing a button, to repeat the last sound or something like it. I shall not attempt to deal with 'pull repeat', the first category. This occurs mainly on seventeenth- and eighteenth-century bracket clocks which are now rare and valuable, and it is very difficult to generalise on the subject, so many are the forms which have been found. A guide to such systems will be found in Harvey and Allix's *Hobson's Choice*.

As we have seen, the repeat facility of the second category – where the clock can repeat what it has last automatically sounded – is available only with the rack system (which was possibly invented for the purpose in or about 1676). Strictly, repeating requires more power and a suitable train, since extra demands are made on the system. However, these are often not provided, save on carriage clocks which to an extent revived the appeal of repetition in a bedroom clock. Thus the usual repeat cord – not to be confused with a cord for letting off the countwheel train to restore correct sequence – is normally just a way of letting off train or trains manually. When this is done, the hour or quarter, now registered by rack and snail, will be sounded. Strictly speaking, such quarter sounding applies only to ting-tang or scale tunes, either striking or chiming, but in practice it is not always so restricted.

Clearly, since the hour rack may register ten, for example, long before and after ten, this can be very inaccurate. In true repeating clocks, therefore, the snail is mounted freely on a starwheel controlled by a spring and jumper and moved forward only immediately before the hour (Fig 12). Thus the chances that the hour repeated is the hour

27

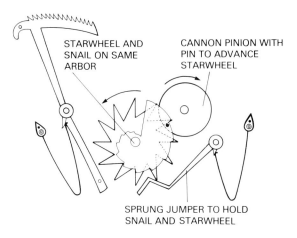

STARWHEEL AND
SNAIL ON SAME
ARBOR

CANNON PINION WITH
PIN TO ADVANCE
STARWHEEL

SPRUNG JUMPER TO HOLD
SNAIL AND STARWHEEL

Fig 12   The starwheel snail

completed and corresponding to the quarters are much enhanced; the
sounding of three quarters and ten must mean three quarters after ten,
whilst without the starwheel snail it could mean three quarters after
eleven. This device is very necessary in repeaters but not elsewhere,
where it was often employed; perhaps it was the latter which prompted
Grimthorpe's typically forthright observation that 'I can see no use for
it [the starwheel snail] and therefore shall describe it no further'. He
went on with the relevant comment that the then (and indeed since)
usual fashion for attaching the cord to the rackhook could lead to
variable results, depending on whether or not the rack was allowed to
fall fully, and suggested instead a lever to raise the lifting piece, which
would automatically release the rack and hold at warning till released.
Carriage clocks, as already mentioned, overcame this problem by
having the repeat lever hold the fly stationary until the rack had fallen;
though even they can be faulted by too hasty a push on the button.
Since British rack-chiming clocks became increasingly lavish (Plates
73–8), and repeating can scarcely have had much functional purpose on
such movements, the fact that they increasingly employed flirt release
rather than a lifting piece perhaps contributed to the fact that
Grimthorpe's suggestion – doubtless in itself not original anyway – was
not widely adopted.

Many repeating clocks have no starwheel snail and many non-
repeating clocks are provided with this device. The French Comtoise
often has a repeat facility, and as these massive clocks were certainly
the clock for an entire household the repeating was undoubtedly
functional in intention, but starwheel snails were seldom fitted.
However, the typical hour-repeating clock of the eighteenth and
nineteenth centuries has a starwheel snail and repetition is by pulling

28

up the rackhook. Normally only rack chimers could properly repeat, but even then there is the problem that the chimes dictated by the chime barrel cannot repeat, but must follow their order. The only way round this difficulty is for the chime to consist of accumulating sequences each of which is the same. Thus, if the quarter rack dictates three sequences, it will not matter which three sequences on the chime barrel are used, since they are all the same. This is why so many chiming clocks are only hour, not quarter, repeaters, and probably explains why apparently boring chimes – ting-tangs and six-bell scales for instance – were so frequently used. Rees says that such scales are the commonest chimes of his day (1750 to 1825 or so), whilst Grimthorpe on the other hand tartly observes that 'this plain distinction' between repeatable and non-repeatable chimes is 'often overlooked'.

Two trains are cheaper and more portable than three; and quarter striking entirely avoids the problem of the non-repeatable chime since the sounding of the quarters is dictated not by a revolving barrel but by the same pinwheel as is used for the hours, striking on one or two bells or gongs according to which hammer tails are presented to it – or obstructed by levers. Thus the quarter striker, with quarter snail and rack and with hour starwheel snail, has been one of the most popular forms of quarter repeater. The vast majority of these clocks strike ting-tang, but four or six bells or gongs have also been used in the more sumptuous specimens. They normally strike the last sounded quarter and the preceding hour as dictated by the starwheel, but other arrangements are found. In ordinary running, the hour alone is generally sounded, whereas in ting-tang chimers the striking of four ting-tangs at the hour is usual. These ting-tang quarter repeaters – grande sonnerie if there is a lever (which can be manually set aside) to allow the hour to sound at the quarter; petite sonnerie if the hour is not sounded – have hour and quarter racks and so can be made to repeat the quarters. They are quite distinct from the less elaborate quarter-striking ting-tangs which have no quarter rack or snail but depend on spaced lifting pins which release from one to three teeth of the single rack at the quarters. This latter type can at best repeat the hours, since the manual release of the rack must be all or nothing.

Strike or Chime Silencing
There are various ways of silencing striking and chiming mechanisms, either as an option built in – an indicator hand working a cam and lever behind the break-arch, or a lever usually close to 3 o'clock – or permanently, as owners, even of valuable and beautiful striking and chiming clocks, sometimes insist. The essential is to adopt a method which risks neither damage to nor stopping of the going train. Ideally

also, the strike or chime will within a short time be sounding correctly when switched back on.

One method is through the countwheel system, where simply leaving it unwound will not damage the clock if it is only a striking clock. However, those who wish for peaceful nights normally require striking as well as chiming to be silenced and, though countwheel chiming can be left unwound, doing this to the usual rack striking may cause damage despite certain safeguards normally built in if the rack is allowed to fall – for example if the chiming stops when the striking has been let off, or in clocks with a separate hour let-off arrangement. In any case, mainsprings are better left wound than unwound. Therefore in many clocks of the chiming 'boom' early in this century it is assumed that the owner will continue winding the piece, but a friction lever which holds all the hammers, or just the chime hammers, away from their gongs is provided. The trains then run as usual and chiming is correct when resumed. Alternatively, a lever raises the lifting piece and holds the chime at warning when, unless it is independently let off, the strike will not run either and its rack cannot fall. In this case, unless self-correction is a feature, the chime will perform immediately when released and the phasing, save by a remarkable coincidence, will be wrong. Therefore – and in case the hands are carelessly advanced – a cord is provided for letting off the chime until its sequence corresponds to the time shown by the hands. The mere removal of a countwheel will not silence a clock permanently; the result will be a one-at-the-hour (or

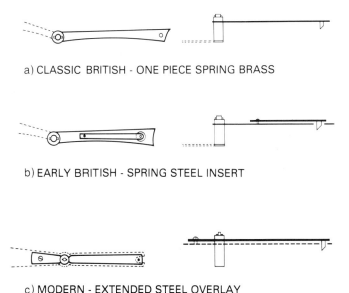

a) CLASSIC BRITISH - ONE PIECE SPRING BRASS

b) EARLY BRITISH - SPRING STEEL INSERT

c) MODERN - EXTENDED STEEL OVERLAY

Fig 13 Safety rack tails

30

quarter) passing blow, and in a countwheel chimer the striking will not usually be let off.

The rack system, as already implied, is more critical in that if the rack falls at twelve and the clock for any reason fails to strike, there is the danger that the rack tail will collide with the vertical face of the snail and, even though rack tails have long been sprung so that in theory they will slide over the snail, there is a real risk of damage and stopping the clock (Fig 13). This is a very common cause of stopping at around one o'clock, and it can often be identified by the fact that the clock comes to life again if the minute hand is revolved anti-clockwise, though invariably the hands and dial at least have to be dismantled to discover why twelve o'clock was not properly struck. All this applies equally to the fourth quarter of rack-chiming systems, when of course the hour will fail to be let off as well.

It is therefore essential to prevent the release of the rack when arranging to silence the clock, and again the most usual arrangement is to raise the lifting piece out of engagement with the lifting pins. Until perhaps the middle of the nineteenth century, lifting pieces in good rack clocks were mounted on a sprung arbor able to move forwards and backwards between the plates. A strong lever with an angled pallet tip bore on the end of the arbor and thus the lifting piece could be brought forwards or backwards, according to the position of the pins, to avoid the lifting pins. (This has also always been the standard means of moving a chiming barrel to produce a change of chimes.) The indicating hand, usually in the break-arch, had various rubrics such as 'Strike/Silent', 'Eight Bells/Silent', 'S[trike]/N[ot strike]' and 'Chime/Silent/Strike', for the device could be applied to striking or to chiming which usually, but not always, included striking. A less common method used a lever to hold up the rack, stopping the pallet tail and train from moving, and this method was also used to hold up the quarter rack to silence rack chiming and striking together. In rack chiming which is let off by a flirt, the usual means of silencing is a lever moved to prevent the flirt from rising high enough to knock out the chime rackhook, which has a pin to catch a notch in the lever (Plate 77). Alternatively, the flirt, like the lifting piece just mentioned, rode on a sprung arbor between plates and could be moved forward out of the path of the lifting pins.

For chiming and striking to be silenced independently, an independent let-off arrangement for the striking is required. Naturally, the striking alone can be silenced by any of the above methods, retaining the chiming, but to give complete peace at night, independence is arranged, somewhat rarely, by means of a twenty-four-hour cam which raises the chime lifting piece. As the cam provides an interruption of set duration, the chiming is correct when resumed. This device is now being revived by a mass producer of high-quality chiming movements.

When, horror of horrors, a customer does demand that a clock be silenced permanently, as far as he or she is concerned all the relevant 'works' may simply be removed. Clearly no restorer can contemplate such a thing, so what is to be done? It is possible simply to bend the hammer so that it fails to strike the bell; but many people will find the whirring worse than the clanging if – as without much confidence one advises – they continue to wind the sounding train. Undoubtedly it is best for all concerned if the sounding trains are not let off. This can be accomplished by wiring the lifting piece out of the path of the pins, or by removing or pushing back the lifting pins. The latter seems the tidiest method, and one which any subsequent surgeon can easily remedy.

This completes our outline of the basic features of sounding systems. In the next chapter we will look at the principles of their gear trains, before considering in more detail how these systems have been used in actual movements and going on to more practical matters. There is a purpose in this order, for the best aid to repair of sounding mechanisms is familiarity with the common principles on which they work and with the wealth of variation which is found.

2

STRIKING AND CHIMING GEAR TRAINS

There is an almost inevitable tendency to think that, because it is not concerned with time, the gear train for the sounding side is more haphazard than that for going. I remember, long ago, proudly furnishing an external countwheel with a wheel and pinion which seemed to fit, hoping all would be well. All was not well. There was no way in which the locking and the action of the countwheel could correspond. The moral was clear: sounding trains require calculation just as much as going ones.

The purpose of the gear train on the going side is a double one. On the one hand, it reduces the relatively rapid vibrations of pendulum or balance to a more useful frequency, such as one revolution per minute, or hour, or day; here the ratios are critical. On the other hand, it distributes over a much longer period the energy put into the raising of the weight or winding of the mainspring which takes only a few seconds; here the ratios are more of convenience.

The purpose of the gear trains on the striking side is similarly dual. They, too, assisted by long intervals imposed by the going, have to spread the input energy over the required going period, the ratios being a matter of convenience. Secondly, they have to maintain certain internal relations so that correct series and sequences of blows are struck. As the number of revolutions in a given period is the criterion for making up the going train, so the exact number of blows and complete sequences is the criterion for the sounding train, corresponding to the 'programme' built in in the form of snail or countwheel. This means that certain gear ratios are no less critical than on the going side. We will therefore look at some of the principles behind different train arrangements, and this will help anyone designing a train or faced with replacing a missing wheel.

There are a number of lists available of standard trains which have been used in the past and, rather than reproduce them here, the reader is referred to the works of Britten, Goodrich, Jendritzki, and Allix and Bonnert. To give the general picture, the following are the names of the wheels in a typical eight-day striking and chiming train; the thirty-hour train is similar, without the intermediate wheel:

	Strike	*Chime*
1	Great, barrel (countwheel)	Great, barrel
2	Intermediate (countwheel)	Intermediate
3	Pin or hammer	Gathering, countwheel (ratio wheel on arbor)
4	Locking or hoop, gathering	Locking
5	Warning (sometimes locking)	Warning
6	Fly	Fly

Five-wheel trains, in which the chime locking takes place on the gathering pallet wheel and where the striking intermediate wheel is the pinwheel, are also common on older clocks.

It will be noted that in countwheel striking there are several options for mounting the countwheel, whose only requirement is a precise relationship in its gear with the pinwheel, as discussed below; it may also be mounted on a stud and driven by a pinion on the arbor of these wheels. In thirty-hour movements, the hammer pins are on the greatwheels; in the rack version, the second wheel carries the gathering pallet – and may do the locking, dependent on the system employed – whilst the third and fourth wheels are warning and fly. Thus, unlike going trains, there may be no wheels concerned solely with the duration of running, although in eight-day clocks the sounding greatwheels or barrels and the intermediate wheels are such unless they carry the countwheel or its pinion.

Basic Requirements

Striking
The locking or hoopwheel revolves once for each blow and there are 78, or 90 with half hours, blows in an hour – 1,872 or 2,160 per day. It follows that to give a reasonable fall of weight or a practicable spring design, a large reduction is required to the greatwheel or barrel, and in practice an intermediate wheel is essential for an eight-day clock, when the barrel usually revolves once in twelve hours. Nothing but convenience governs this overall ratio but, within it, it is essential that the ratio from locking wheel to countwheel be 1:78 or 1:90, since the countwheel normally revolves once in twelve hours. It is also necessary that the ratio of the locking-wheel pinion to the teeth on the pinwheel which drives it be the same as the number of pins on the pinwheel, since the locking wheel must revolve once per blow or pin. Thus the number of pins on the pinwheel, which depends in part on the design of

the hammer tail and space required for its movement, effectively determines the ratio of pinwheel to locking-wheel pinion. So we commonly find on early British clocks a locking-wheel pinion of 7 and a 56-toothed pinwheel with 8 pins or, on a thirty-hour clock, a hoopwheel pinion of 6 (like most thirty-hour clock pinions) driven by a greatwheel of 78 teeth and with 13 pins. Whatever the ratio chosen, it must be an integer. This is a golden rule of sounding trains, since wheels are required to stop repeatedly in the same position.

The third requirement is that the ratio of the locking-wheel teeth to the warning-wheel pinion must also be an integer; what ratio is immaterial, except to the speed of striking. Thus, for example, we may have a locking wheel of 60 teeth and a warning pinion of 6. The reason is of course that locking and warning must always occur in the same place relative to each other; ie for a given locking position, the warning pin must reach the warning piece after the same interval on each occasion when locking is released. And this rule is normally followed also for clocks without warning, where the position of the wheels after locking is, in fact, immaterial.

Fourthly, the countwheel must revolve once in twelve, rarely twenty-four, hours. How this overall reduction from 78 or 90 of the hoopwheel or locking wheel is achieved varies, but there are a number of standard trains, involving 4, 6 and 8 leaf pinions driving external countwheels. As we have seen, a reduction of 1:13 has usually been achieved by the greatwheel, and the problem is then to find the combination of pinion and countwheel gear which will bring about the necessary remaining reduction of 1:6. Where the pinion is of 4, it commonly consists of 4 pins carved out of the greatwheel arbor and meshes with a 24-toothed wheel with very large teeth to which the countwheel is riveted. Otherwise, the greatwheel arbor is squared to carry the pinion, which is covered and held in place by the external countwheel. It was the fashion with pin-countwheels to use pinions of high count, resulting in very large gears and countwheels covering much of the back (Plate 10). Where the old countwheel is internal – confined largely to eight-day clocks, probably because it was so convenient to have the barrel revolve once a day – it is attached to the barrel, usually with spacers (Plate 8). In many nineteenth-century American and German movements the internal countwheel is mounted freely on a stud, or loose on the pinwheel arbor, and has ratchet teeth cut round its circumference so that one tooth corresponds to one of the 78 or 90 divisions. These teeth are caught by a pin extended from the hoopwheel pinion, which advances the countwheel one tooth per blow; thus the required 1:78 reduction is achieved.

In the month movements which were a luxury of the seventeenth and eighteenth centuries, countwheel trains were usually curtailed by using a hoopwheel with two opposite slots and a multi-pinned pin-wheel, with the countwheel mounted on the intermediate wheel arbor. French grande-sonnerie carriage clocks adopt the same principle, using a double gathering pallet and two pins on the locking wheel. This in effect reduces the required ratios by half, so that the overall reduction to the countwheel need be only 1:39. The example in Plates 36–9 has the train

$$\frac{\text{hoop pinion 6}}{\text{pinwheel 72}} \times \frac{\text{external pinion 16}}{\text{countwheel 52}} = \frac{39}{1}$$
$$\text{(24 pins)}$$

In the common round French countwheel movements, the count-wheel is mounted on a squared extension of the intermediate wheel arbor (Plates 15–16), and related to the locking wheel 1:90. The same rule is followed, and typically there are 10 pins on the pinwheel, which here follows the intermediate wheel. With a locking-wheel pinion of 7, and 70 teeth on the pinwheel, the ratio (10:1) is thus, as usual, equal to the number of hammer pins. The same principles apply to rack movements, except that there is no equivalent to the countwheel-locking ratio.

The ratio of the remaining parts, ie the warning wheel and fly, is purely a matter of preference, being a main influence on the speed of sounding (page 38).

Chiming

The requirements of chiming trains are similar to those for striking. For example, the relation of chime locking wheel and countwheel in countwheel chiming must equal the number of divisions in the countwheel – 78 for the hour but only 10 for the quarter countwheel – so that the train can be locked after the smallest unit, either blow or sequence, has sounded (at a quarter past). Again, there must be an integral relationship between locking and warning wheels. But of course the notion of revolutions per blow is now replaced by revolutions per sequence, and we must ensure that the train stops and starts exactly between sequences as measured by the detent and countwheel or by the rack tail and snail. With the countwheel, one of its ten cumulative divisions corresponds to a sequence, and so does one tooth of the quarter rack and one step of its snail.

As we saw in Chapter 1, in the vast majority of cases a chime over four quarters is composed of five sequences repeated, the first sequence occurring for the second time at the end of the third quarter; although with a short chime it is possible for all ten sequences to be pinned round the barrel (Plates 57–8). It follows that the chime barrel which

36

programmes these five sequences must revolve in a ratio of 2:1 to the countwheel – which revolves only once in an hour and on whose arbor the barrel could otherwise be placed direct – and 1:5 to the gathering pallet which revolves once for each sequence. These ratios can be achieved in several ways. Generally, until the end of the nineteenth century, they were worked out internally and the chime barrel, if not too long, was between plates. The countwheel was usually outside the front plate, as it continued to be in the first part of the present century; but there are many late nineteenth-century clocks, often of German provenance, with internal countwheels, anticipating the growing fashion of placing the chime barrel outside the back plate. The internal countwheel was mounted on the intermediate wheel's arbor, while the greatwheel revolved once in twelve hours and engaged the countwheel in a ratio of 12:1, giving typically 96 teeth on the greatwheel and a pinion of 8 for the intermediate and countwheels. The internal barrel was driven, as it were, at a tangent to the main train, being taken off the intermediate wheel in a ratio of 2:1. Since its positioning is so critical, and independent of the main train, taking the barrel outside the back plate was a logical development which much simplified assembly and adjustment.

The twentieth-century external countwheel is fixed by a set-screw to a forward extension of the third-wheel arbor which is extended to the rear to carry an adjustable wheel which meshes, with or without a third wheel, with the chime-barrel wheel – again in a ratio of 2:1. The constituents of this ratio are of no importance. So long as they double the speed, the wheels are required merely to convey power to the somewhat removed barrel on the back plate or below it (underslung). Sometimes there are pairs of identical wheels (ratio wheels) with adjustable set-screws, which are a great advantage in that the barrel can be positioned without dismantling the movement.

Just as the countwheel revolves at half the speed of the chime barrel, so the rack system's gathering pallet has to revolve five times as fast as the barrel. (The barrel revolves twice an hour, whilst the gathering pallet revolves once for each of the ten sequences.) The arrangement has usually been to make the ratio of the intermediate wheel to the gathering-wheel pinion 10:1, and then to drive the barrel from the intermediate wheel in a ratio of 2:1. This can be done equally well internally or externally by the use of ratio wheels, which were not uncommon, from the mid-eighteenth century, in the relatively lavish clocks to which the full rack system is suited, and which were necessary with barrels extending behind or parallel with the movement because their carrying a choice of chimes made them long. However, all that is required for a basic internal arrangement is for the chime-barrel's wheel to be half the size of the intermediate wheel.

As with striking systems, the ratio of warning wheel to fly is not critical, but it plays an important part in controlling the speed of chiming.

Speed of Sounding

The speed of sounding is determined by many factors including energy put in, friction encountered, weight of hammers and the strength of their springs; and also by the gearing, from locking wheel on. As time has passed, gongs have superseded bells, partly on economic grounds and partly because gongs of any kind have a more lingering sound. Therefore higher warning wheel to fly ratios have tended to become popular. But there has also been increased attention to the adverse effect of mass at the fast end of the train so that, as noted already, the small and heavy flies so common throughout the eighteenth century and earlier, have given way to larger, lighter, more complex, even variable, devices. Similarly, it has become usual to make the warning wheel and fly pinion in a smaller gear module, ie with smaller teeth, than that of the remainder of the train, thus maintaining a high ratio without large mass at the end of the train. Similarly, fly pinions have always tended to have too low a count from the point of view of strength and wear, and are often provided with adjustable bearings to offset the effects of the latter.

Since speed of striking is very much a subjective judgement, a good deal can be done to the various springs to alter speed without radically modifying a clock. It seems clear that the old makers preferred a faster speed of hour striking on bells than we find entirely to our taste; in fact most people find the very slow speeds adopted on some 'directors' clock' gongs quite acceptable. There is no justification for altering the known or likely character of an old fly or for putting in a higher count warning wheel. If warning wheel and fly are missing, however, it may not be possible to be sure of the gear size involved or, therefore, of the counts which will fit the available depth. In this case, experiment may be needed both with the count and with the size and shape of the fly; for instance whether or not to have rounded corners, which seem to occur at least throughout the eighteenth century although square-cut ones are usual. So far as striking goes, the practical extremes of warning wheel to fly ratio seem to be 9:1 (fast) and 7:1 (slow). Chime ratios are very much higher and there is a greater range – 13:1 (wheel 78, pinion 6) seems to be a favourite of the rod-chimers of the 1920s and 1930s, and I have found 15:1 (90, 6) in an Edwardian rack movement with massive, resonant coiled gongs. On the other hand, with rod gongs, ratios as low as 10:1 are found.

Some examples of the fast ends of sounding trains of some of the movements illustrated in this book are set out in the table opposite. It

	1 Hoop/Locking/ Pallet wheel (once per blow or sequence)	2 Warning pinion	3 Warning wheel	4 Fly pinion	5 Warning: fly ratio	'S'
Month striking, c1700 (bell)	72 (2×36)	6	56	6	9.3	56
Thirty-hour striking, c1690–1710 (bell)	60	6	48	6	8.0	80
Internal-rack striking, c1740 (bell)	54	6	48	6	8.0	72
Pin-countwheel striking, c1770 (bell)	48	6	48	6	8.0	64
Carriage striking, c1880 (gong)	72	6	72	6	12.0	144
French countwheel striking, c1870 (bell)	60	6	60	6	10.0	100
Ting-tang, c1760:						
striking (bell)	49	7	42	7	6.0	42
chiming (bells)	49	7	42	7	6.0	42
Modern countwheel:						
striking (rods)	66	6	60	6	10.0	110
chiming (rods)	60	6	72	6	12.0	120
Modern countwheel:						
striking (rods)	60	6	60	6	10.0	100
chiming (rods)	66	6	72	6	12.0	132
*Standard longcase striking (bell)	49	7	48	7	6.9	48
*Standard bracket:						
striking (bell)	70	7	60	7	8.6	86
chiming (bells)	64	8	50	8	6.3	50

*Taken from Britten, *Horological Hints and Helps* and *Watch and Clockmakers' Handbook*. The bracket-chiming ratio, however, seems untypically low.

is true that the many other factors involved make comparison between different types of clocks of limited value; C. Allix's comparisons of carriage-clock trains are more revealing. However, some points do emerge, such as the low ratios found in weight-driven bell-sounding clocks and also in bracket clocks of corresponding period and bells; the high ratio for the little French movement of perhaps 1880 with its bell; and the still higher ratio for the gong-sounding carriage clock. These are trends which it is helpful for a restorer to know, when faced with four empty holes in the plates.

Sample Wheel Counts
Indication of relative speed is given by

$$\frac{\text{Column 1}}{\text{Column 2}} \times \frac{\text{Column 3}}{\text{Column 4}} = \quad \text{fly revolutions per blow of strike or sequence of chime ('S').}$$

3

COUNTWHEEL STRIKING

The external countwheel is probably the oldest form of striking and in thirty-hour clocks it reigned supreme from the early seventeenth century – when house clocks, in the form of lantern clocks, seem to have become reasonably normal fittings of better-class households – until 1780 or even later in country districts. It was, however, outstripped by rack striking in quality eight-day longcase clocks from about 1725, and very much earlier in bracket clocks where repeating work was appreciated. Throughout the two centuries of its greatest popularity it continued in very much the same form (Plates 1–4), with materials inherited from lantern-clock practice, and this continued later in cottage thirty-hour pieces (Plate 5).

Both these clocks have a thirty-hour longcase movement of the posted construction known as 'bird-cage', with the striking train placed behind the going. Rapidly, probably under London influence and the fashions for rack striking and key winding, such construction was superseded in the eighteenth century by full vertical front and back plates with the trains alongside each other, whether or not with endless chains or key-wind. At first, in thirty-hour and other countwheel mechanisms, the striking train was on the right; but internal countwheels – seldom found on thirty-hour clocks – and rack systems, led to the adoption of the left as the standard position for the striking train.

The ring-and-punch ornament of the dial in Plate 1 shows it to be of Northern England origin, although the engraved sheet spandrels are very unusual anywhere; and the pallet shapes indicate a date of around 1700. Nevertheless, it has obviously much in common with the second, later, movement (Plate 5), which represents continued tradition. In each example, the very large bells are typically mounted on substantial standards screwed to the top plates, and the hammers swing from right to left inside the bells. Typical is the stout, L-shaped, iron hammer stop descending from the top plate, and the equally hefty iron hammer spring which could be mounted either on the top or on the bottom plate. There are two arbors on the right of the movements carrying respectively the lifting and warning pieces, and the link to the locking piece, all three on the same arbor (Fig 14), although the link was

Plate 1   Dial of English thirty-hour longcase movement, c1700
Plate 2   Hoopwheel and countwheel detents, warning piece, and link piece of the longcase movement (Plate 1)

sometimes attached to the lower arbor and the locking piece and detent. These arbors until quite late in the eighteenth century tend to be of square section and the various projections from them are brazed into slots. On the upper arbor, the link piece, locking piece and countwheel detent could be a single piece of iron brazed on, although on later movements the detent, like the lifting piece, is usually pinned to the squared end of its arbor and so is completely outside the back plate. In these thirty-hour clocks the strike greatwheel is also the pinwheel and, as we have seen, has normally thirteen pins to act on the hammer tail. The hammers are very substantial and are pivoted on square arbors to the left of the pinwheel.

The second movement (Plate 5) is a provincial example from the mid-eighteenth century, as shown by its sheet pallets and in particular by the fact that it has two hands. This affects the lifting arrangements. The single-handed movement (Plates 1–4) has a large hook-shaped lifting piece which rises and falls on a twelve-toothed starwheel attached to the hour wheel which, being attached to the dial and merely resting in the front-plate hole, is not in the illustration. The hour wheel is driven directly by the large pinion at the front on the

Plate 3   Hammer spring and stop of the longcase movement (Plate 1)
Plate 4   Typical lifting piece for twelve-hour starwheel (not shown) and single hand of the longcase movement (Plate 1)

going greatwheel arbor. However accurately the starwheel is cut, this rather crude and large-scale set-up always seems to result in striking which is reliable only to within some five minutes – quite sufficient at the time, but a little inconvenient in the modern household. Where there are two hands, the lifting is by a pin on the minute motion wheel (anti-clockwise) in accordance with virtually universal British practice from then until modern times.

A summary of the action of both movements illustrates the basis of countwheel systems. When the lifting piece is raised, the warning piece on its arbor comes into the path of the pin on the warning wheel. The latter does not immediately move; there is a little clearance between the link piece and the warning piece to ensure that the latter is truly in the path of the pin before the locking piece rises from the hoop slot and releases the train, also raising the countwheel detent on its arbor out of the countwheel slot. Once the train is freed, it runs the short distance allowed until the pin on the warning wheel meets the warning piece, when the train stops, at the ready. It can be freed now only by the release of the warning pin, accomplished by the warning piece falling back when the lifting piece drops off the lifting pin or starwheel tooth.

It is clear that the whole system depends on the hoopwheel's locking (which can occur at a single blow) coinciding with the raised sections of the countwheel which prevent this locking so long as they are obstructing the fall of the countwheel detent and the locking piece on its arbor. It should also be borne in mind that the hammer is heavy and its spring very strong – strong enough to slow or even stop the train until it has gained momentum. Further, the run of the warning pin until it meets the warning piece must allow the detent complete freedom of the countwheel, so that it can fall next only on a raised section. Therefore these trains have to be assembled in a certain way, the wheels being put into the plates in such a manner that the warning wheel has about half a revolution's run with the locking piece in the hoopwheel slot. The countwheel can then normally be placed on the outside in a position where the detent is in a countwheel slot. Although this sounds straightforward, the hammer spring cannot always be inserted later and, if it cannot, it has a tendency to turn the wheels during assembly. Thus the assembled movement must often be adjusted by raising the plates to the extent that the warning and hoop wheel can be turned relative to each other until the proper position is reached. There is the further complication that the hammer tail should not be engaged with a pin on the pinwheel when the train is locked or,

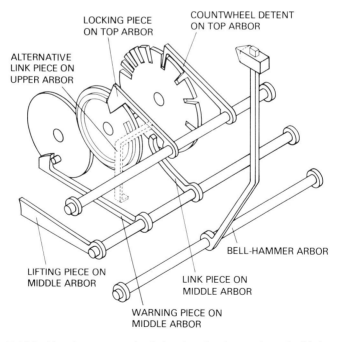

Fig 14   British thirty-hour countwheel showing the three arbors (in 'bird-cage' movements the hammer arbor is to the left, on the other side of the movement)

44

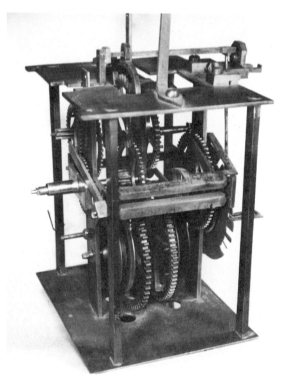

Plate 5   Provincial mid eighteenth-century thirty-hour longcase movement with lifting
piece on minute wheel (two hands) and typical traditional striking work

ideally, when it is at warning; the train should be well under way
before the burden of the hammer is taken up. Depending on the
clearance of the hammer tail, however, and on the amount of 'run to
warning' which is allowed, it may be unavoidable that the hammer is
engaged at warning, though it must never be so when the train is
locked.

Many of these arrangements apply equally to more sophisticated
clocks with external countwheels. Plates 6–7 show a London-made
movement of about 1700 which goes for a month; the use of the
double-notched hoopwheel and the driving of the countwheel from the
intermediate wheel, with a resultant high position, have already been
noted in connection with gear trains. The posted frame is replaced by
vertical plates, but the striking is still on the right. The bell hammer
strikes the bell on the outside and is central on an arbor between the
plates, being fitted with a separate stop piece; and the long hammer
spring attached to the back plate was typical of British movements for
two centuries and more. The lifting is from the minute wheel but many
differences from the previous movements' details can be seen. The
lifting piece is solid with the warning piece in a forked design, since the
warning has no longer to pass through the movement to a train at the

Plate 6   Month movement by Robert Williamson, London c1700, with lifting, warning and link pieces and hoop locking piece

Plate 7   Manual let-off, hammer stop, warning piece and typically high position of countwheel on the Williamson movement (Plate 6). The bell standard is (rather unusually) fixed to the inside of the back plate

rear, and is bent round through a slot in the front plate to engage the warning wheel. Its arbor is provided with a protruding hook by which the striking can be let off and adjusted manually. The link piece and locking are taken to the rear of the movement and the countwheel detent is pinned externally to the extended locking-piece arbor. The bell is mounted on the inside of the back plate; earlier, it might have been mounted on the front plate; later it was more often mounted on the outside of the back plate, often rather awkwardly within the pallet bridge (Plate 11). The sequence of events is, however, precisely as in the thirty-hour pieces already considered and the same points arise in assembly with the difference, typical of most later movements, that in setting up the train the exact position of the warning piece cannot be judged so easily as when it is between plates.

For a fairly short period of about forty years in the first part of the eighteenth century, a fashion for the internal countwheel arose naturally from the fact that the greatwheel could revolve in twelve hours in eight-day clocks. Instances in thirty-hour clocks are extremely

46

rare. Earlier clocks used an external countwheel on a rear extension of the greatwheel arbor, which was in fact split for this purpose (Fig 15) so that the countwheel was in effect attached to the greatwheel rather than to the winding square. In the Liverpool clock of about 1720 illustrated in Plates 8–9, the countwheel is, as became general, attached with spacers to the greatwheel which is of course divorced from the arbor by click-work for winding. It will be seen that the striking has moved to the left, as is universal with this type of countwheel. The bell hammer now has a combined stop and spring; the stop being the heeled end to the spring, a device which in various shapes became very general. The bell is mounted on the outside of the back plate, to the left of the pallet bridge. This is a high-quality movement: the pillars are very nicely turned, the fly is rounded, the striking levers are of polished steel. As the striking is on the left, the lifting is taken from a pin on the cannon pinion, not the minute wheel. This was to become standard practice for countwheel movements once striking on the left was established, whilst rack movements normally lift on the right from the minute wheel with a cranked lifting and warning lever. Here the old form of internal warning piece (Plate 9) is adopted, whilst the low position of the countwheel and relatively high position of the hoopwheel necessitate a long countwheel detent curving down from the hoopwheel arbor. In assembly there is the additional problem that no variation of the countwheel relative to the train is possible. The whole train must be assembled with wheels correctly positioned, the locking piece in the hoop slot and the detent in a slot of

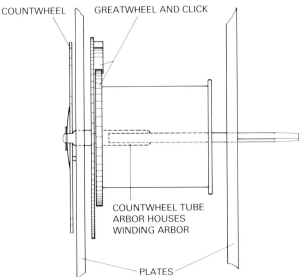

COUNTWHEEL   GREATWHEEL AND CLICK

COUNTWHEEL TUBE
ARBOR HOUSES
WINDING ARBOR

PLATES

Fig 15   Early split greatwheel arbor, external countwheel (eight-day)

47

Plate 8 Eight-day internal-countwheel longcase movement by Isaac Hadween, Liverpool *c*1725. Note lifting piece from cannon pinion on left, and long countwheel detent from the top arbor which also carries the hoop locking piece. Typical one-piece hammer spring and stop

Plate 9 Note internal warning piece on middle (lifting-piece) arbor of the Hadween movement (Plate 8)

the countwheel, and this must be done with the hammer spring, deep within the movement, in position.

In the second half of the eighteenth century a rather different type of countwheel developed, particularly in the North of England, and became popular on thirty-hour cottage clocks. Countwheels with projecting, rather than raised, sections, seem to have existed at all periods and their working is no different; however, the pin-countwheel now to be examined works on the reverse basis – the train is locked when the detent is up, on a pin, and free to run when it is down, between pins. Plates 10–12 illustrate such a movement, dating from about 1775. The forked lifting and warning piece is similar to that on the month movement (Plate 4) and the bent-back warning piece is particularly clear, although having the lifting pin behind the minute wheel is a curious detail. The train, typically still of a thirty-hour movement, is on the right, but it will be noted that instead of the usual three arbors (hammer, lifting, locking) there is only the hammer arbor and one other, which is attached to the countwheel detent, locking

Plate 10   Front view of thirty-hour pin-countwheel longcase movement, c1775
Plate 11   Note the single (top) arbor which carries lifting, warning and locking pieces of the movement in Plate 10. The train is locked with the countwheel and detent in this position.

Plate 12   Side view of the movement in Plate 10, showing cut-out fly (to clear the pillar), conventional hammer spring and the blade of the locking piece (pin locking)

piece and lifting/warning piece. No doubt this economy was one reason for this type of clock's popularity. The explanation is that the usual clearance between warning and unlocking – the justification of having an adjustable link between warning and locking arbors – is built in and fixed. The train locks against a pin, not in a hoop, and this pin locking was to become very much more general after this time in clocks of all types. The warning, locking and detent pieces are (or must be) so placed on their arbor that unlocking cannot occur before the warning pin is in position, at which time the hooked countwheel detent will be well clear above the countwheel pin (Fig 16). The action of the locking piece, uniquely to this system, is double, being above and below the locking pin on which it catches when at rest. At warning, the lifting piece raises it above the locking wheel pin whereas, when the train runs, it falls below, as dictated by the countwheel. The countwheel detent is so shaped that, as a pin on the countwheel encounters it, the detent and locking piece rise until the locking piece meets its pin and stops the train, when the extremity of the detent hook rests on a countwheel pin.

These movements can be awkward customers. The chief problems are the arbor with its three pieces, and the countwheel. The depth of engagement of the locking relative to the height of the countwheel detent is critical and is affected by a bent or loose locking piece, a worn or bent countwheel detent, and worn or bent pins on the countwheel. With so many variables it is difficult to know where to start, but probably trueing the countwheel pins is the prerequisite. One can then go on to ensure that the locking pin is just below the top edge of the

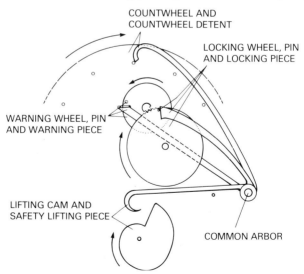

Fig 16 Diagram of the pin-countwheel arrangement, with all pieces on the same arbor (see Plate 11)

50

Plate 13 American Waterbury Clock Company internal-countwheel, thirty-hour weight-driven wall movement, c1870. Note extreme economy in use of metal and typical external placement of pallets and escape wheel

Plate 14 View of the Waterbury movement (Plate 13). Hook on fly arbor for warning, hoopwheel catching from the rear, free-running countwheel driven by a hoopwheel pinion leaf (not visible) and typical combination of wheel teeth and countwheel sections

locking piece when the detent hook is almost as high as it can go on a countwheel pin. If it goes too high, or locks too shallowly, the clock may strike at warning, and it is well to keep the run to warning down to about a quarter of a revolution. If the clock fails to strike when warning is released, it is likely that the detent is still catching a countwheel pin, in which case relocate the countwheel relative to its pinion.

Except in chiming, the countwheel, as we have seen, suffered something of a decline in British clocks after the eighteenth century, but it continued in France, though superseded by the rack in carriage clocks, for at least another century. And it was the basic system adopted for real mass-production of clocks beginning in Germany, probably influenced by the Black Forest tradition, and developed widely in the USA in the mid-nineteenth century.

The American shelf and wall clocks mass-produced by firms such as Waterbury, Ansonia and Holloway are at first glance somewhat forbidding with their skeletonised plates, compact arrangements and apparent mass of steel wires in place of solid British levers. There were variations but, in general, internal countwheels were preferred for striking and chiming, and the form these countwheels took highlights the emphasis on economy. Rather than, in British fashion, being a heavy brass disc with its associated wheels riveted to it, the typical countwheel was in fact a large gearwheel with 78 or 90 teeth (representing the usual divisions) and slots, each representing one tooth, cut round. Into these fell a flattened wire detent on the same arbor as a wire locking piece which caught from the rear – rather than met from the front as we have seen so far – a brass notched cam corresponding to the British hoopwheel and responsible for locking. The second arbor contained, as usual, the lifting 'and warning piece, likewise of wire, and the link to the first arbor. Sometimes lifting was accomplished from a cam attached to the cannon pinion, and some-times, as in the Waterbury thirty-hour clock in Plates 13–14, even the lifting was internal, there being a 'flag' projecting from the central arbor like a large pallet, which raised the wire lifting piece. As we have already noted, a popular mounting for the countwheel was on a stud in the front plate or free on the pinwheel arbor, where it was simply advanced tooth by tooth at each revolution of the locking wheel by means of an extended pinion trundle on the latter.

The escapements of these movements, which are mounted on the front outside the plate, are rather prone to wear; but the striking mechanism, for all its spider-like appearance, is remarkably efficient and well designed – for instance, the large, light fly employed is noteworthy. The clocks strike on coiled-tape gongs, which probably originated in Black Forest clocks and reached Britain and the USA in the middle of the nineteenth century. Early examples of mass-produced

Plate 15 French eight-day external-countwheel table clock movement, *c*1880. Note typically combined locking piece (pin) and countwheel detent, also curved piece linking warning to the locking arbor. The lifting/warning piece is typical of French countwheel and rack movements

Plate 16 Rear view of the French movement (Plate 15) showing combined locking piece and detent and typical shape of locking piece against pin. The spoked countwheel with sloped exits to the stops is also typical, as are hammer and bell standard, and the high finish throughout. The separate cock for setting the hammer tail off the pinwheel pins after assembly is clearly visible

clocks employ trains with ratios not generally very different from those used with bells, and consequently strike very much on the fast side for modern taste. Later, higher ratios became more common. Again, whilst later clocks with gongs have the hammer heads faced with leather or even rubber to mute the clangour, this was apparently part of the distinctive appeal of early clocks with gong, and their hammers are unfaced or faced with lead as in the Waterbury clock illustrated.

The Waterbury clock has no warning wheel. The large fly is necessary both to reduce the speed of strike and also because it is on the fly that the warning piece acts, by engaging with a short wire hook alongside the fly. This again is for the sake of economy. Even the symmetry of the movement illustrates this – the lantern pinions are all of the same count and, including the countwheel, there are no less than five identical 78-toothed wheels whilst the countwheel, though ratchet-toothed, could be cut from the same stock. These clocks have a certain beauty and they were not made, consciously or unconsciously, to last as antiques, but the cost-cutting does have its drawbacks, particularly in the spring-driven versions. The holes often need bushing throughout and the lantern trundles may need replacement, both of which are quite time-consuming jobs in view of the relatively low value of the clocks even now. The great danger on the striking side is bent wires; efficient locking and counting, with the detent falling radially into the middle of the countwheel slots, are essential. Whether the warning is on warning wheel or fly, there is no clearance between the lifting-lever link and the countwheel detent with which it engages. With a warning wheel a good half turn to warning is needed, and it may be necessary with either system to bend the warning piece up slightly to ensure that it catches first time round.

French mass-produced movements were the product of quite different methods, and high quality resulted – almost a different philosophy seems to distinguish the clocks in Plates 13 and 15. Movements such as that in the latter Plate were made throughout the later eighteenth and the nineteenth centuries and are difficult to date; this one perhaps dates to about 1880. The parts would probably have been mass-produced in standard form and then assembled at one of several centres. Here we make a return to the external countwheel with pinwheel locking as hoopwheel or cam locking becomes virtually restricted to chiming mechanisms, where it is made adjustable with a set-screw.

The typical French arrangement includes many details already seen in earlier movements. The left-hand striking and lifting piece activated by the cannon pinion recall the British internal countwheel, but here the warning piece with its blade piercing the front plate is found on the left, whereas on the British movement there was a separate warning piece on the lifting-piece arbor. Here the countwheel rides on the intermediate (pin) wheel arbor. It must be said, however, that this comparison is with British practice of perhaps 150 years earlier. Contemporary British striking comparable with this French movement would be hard to find, for Britain almost without exception used the rack system by then, with the warning piece on a stud to the right. However, the two lifting and warning pieces are of similar shape, although in reverse positions. In the French movement the locking arbor has still to be connected to the lifting piece, and this is done by means of a curved detent above the warning piece on the front plate, which has the locking piece and detent on its arbor. This link piece is usually, but not always, assisted by a leaf spring in holding down the lifting assembly. The detent and locking piece are a single branched piece of steel, the detent branch extending through a notch in the back plate to the countwheel.

It will be recalled that with the pin-countwheel (Plates 10–12) the detent ended in a reversed hook, pressure against which would cause it to rise and bring about locking. This is because, unlike the sloped face

Plate 17   Thirty-hour Black Forest external-countwheel cuckoo clock, *c*1890. Note wire loop from the brass locking piece round to the countwheel detent; also the typical lifting piece. The action of the bird by the locking piece is clearly visible, as are two of the three 'hammer' tails and the simple stop for the gong hammer itself

Plate 18   Side view of the Black Forest cuckoo clock (Plate 17) showing the wire link piece from lifting piece arbor to locking arbor, and the typically projecting sections of the countwheel

Plate 19   Lifting piece (on cannon pinion pins) of the Black Forest cuckoo clock (Plate 17) and operation of the bird

Plate 20   Rear view of the Black Forest cuckoo clock (Plate 17) showing typical securing of parts by wire hooks, the projecting countwheel, the three 'sound' linkages and the incomplete stopwork

of the hoopwheel detent of older type clocks, there is no lift available from the locking. The same is true of the French pin locking, where invariably the lift is given by a sloped leading edge to the countwheel slots. These are wide relative to the slots of a British countwheel and this, of course, is because French movements always strike one at the half hour – the spaces represent the twelve extra blows needed in a day. Also typically French in Plate 16 are the adjustable hole for the fly and the separate cock for the pinwheel; the latter facilitates setting up the movement correctly with the hammer properly off the hammer pins, and the position of the locking wheel in the correct relationship to the warning wheel. Adjacent wheels (notably pinwheel and locking wheel) in these clocks are often marked with a punch-mark between teeth, and a filed-off pinion leaf, to show the intended engagement; and these should not be confused with later repairers' dots and dashes. Attention to detail, even down to the ornamental rings on the hammer head – which is for a bell, being metal rather than leather-faced – is a feature of these marvellously fine and well-finished movements. They run on remarkably little power, provide few problems in assembly save the occasional broken pivot if care is not shown, and, except for the vulnerable centre-wheel pivot, are often almost free from wear as a result of scrupulous hardening and polishing.

The Black Forest clock industry, long isolated from developments elsewhere yet later having remarkable influence on them, for centuries produced, with little basic change, wooden-framed movements with external countwheels. At first they had wooden arbors and wheels, but over the years more and more brass was introduced. The thirty-hour cuckoo clock in Plates 17–20, a spring-driven model of about 1890, is a late example. The striking is typically on the right and the usual line up of arbors is present. The three lower ones bear, from top to bottom, the gong hammer and the linkages to the two pipes. The lifting piece is hooked, in the manner of a thirty-hour one-handed clock, and raised by two pins on the cannon pinion. To it, towards the back, is attached a short wire lever which raises the hoopwheel locking piece on the arbor above, from which stands up a cranked wire detent passing out between wooden 'plate' and pillar to the countwheel (Fig 17). The lifting and locking arbors are the two at the top.

The Black Forest system is distinctive in that locking and warning are performed on the 'warning' wheel; but there is also a cam or hoopwheel which coincides with the countwheel and determines whether or not locking can occur. The brass locking piece on the topmost arbor (there are five in all) also has a hook to catch the vertical face of the hoop or cam and a tip which catches the warning and locking pin, which cannot engage unless the locking piece has fallen into the hoop. The sequence is therefore that the lifting piece, by means of the warning-piece lever, releases the train so that the warning pin drops off the locking piece and onto the warning-piece tip. This is accomplished without a full removal of the locking piece from the hoop. The lifting piece continues to rise until the locking piece is well clear of the hoop, when the train is held solely by the warning piece and runs when lifting and warning pieces have completed their rise.

The bird's hinged wings are opened, and its tail made to rise and fall, by a wire link from the lower-noted bellows. The doors are linked to the bird by wires and its forward movement opens them. The pipes and gong are operated by hammer tails and levers, with a pin of the pinwheel clear between them. As is usual in older Black Forest clocks, the countwheel has projecting sections – forming the wheel from a thick blank in the lathe, boring it and cutting out the sections was work well suited to production in the home. The countwheel has its own teeth and rides on a stud, being driven by a pinion on the intermediate wheel arbor (the pinwheel). The example illustrated, like many of its time, had simple stopwork for the mainsprings. The pinions at the back are missing, but the blank teeth in the wheels providing the stop can be

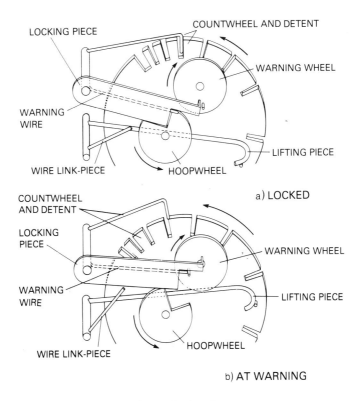

Fig 17   Countwheel cuckoo action

58

Plate 21 Typical thirty-hour Black Forest movement, *c*1890, with similar lifting and locking arrangements to clock in Plates 17-20

Plate 22 Side and rear view of the Black Forest movement (Plate 21) showing curious path of hammer wire and long downward spring to secure upward striking of the gong. Construction, countwheel and locking (as in cuckoo movement, Plates 17-20) are all characteristic

seen. The mechanism of modern cuckoo clocks is almost entirely different (see page 91).

Very near in date is the simple Black Forest thirty-hour striker illustrated in Plates 21–2. This is not a cuckoo clock, but it employs exactly the same striking mechanism and is very similar in detail. Its curiosity is its hammer action. Most striking devices take advantage of gravity, assisted by springs, to direct the hammer downwards onto the gong. In this movement, by a most circuitous arrangement, the action of the hammer is upwards and the vertical hammer spring keeps the hammer up at rest against its pin stop.

These clocks are more difficult to repair than they should be. There is often difficulty with wire linkages which, if they catch or are too long, stop the striking. The double-locking arrangement and run to warning are difficult to set because of the small depth of most movements and the fact that the wooden frames do not reflect light. Many fittings are by wire hooks pressed into the wood; in time these holes become enlarged and have to be plugged. The brass pivot bushings wear and often come crooked in their bores. The lantern pinions, too, are prone to wear and break. One should never be beguiled by the rustic look of a Black Forest clock into thinking its repair will be straightforward. The striking mechanism is not easy to illustrate and should be studied in action, as far as it can be, before the clock is dismantled, suitable sketches being made of its behaviour.

4

RACK STRIKING

Despite the not very conclusive statement of William Derham in his *Artificial Clockmaker* (1696) that the rack system was invented by Father Edward Barlow, a Roman Catholic priest whose real name was Booth, in 1676, the actual origin of rack striking is obscure – lost in the many attempts at that time to find a satisfactory system of (pull) repeat striking. Barlow did indeed apply for a patent for this in 1686, and he discussed it with Tompion who quickly adopted it. But whatever its origins, the system, for striking on the run, did not gain general acceptance for another forty years. Thus, reputedly, its invention was to make repetition possible – clearly a desirable facility when hands could not be seen at night – and certainly it did make possible the wide variety of repeat devices found in the seventeenth and eighteenth centuries. However, it was also widely used where no repeat arrangements were made. As we saw in Chapter 3, it became the normal mechanism for quality longcase and bracket clocks during the eighteenth century, being more expensive and elaborate to make and more prone to wear than the countwheel system, but more reliable and convenient. But it never completely ousted the countwheel, even in thirty-hour clocks, or was adopted so universally in chiming systems. The countwheel was, in the nineteenth century, probably more popular on the European Continent and in the USA than in Britain; but it was revived in the latter as the standard chime-control arrangement for the mass-produced chiming clocks of the 1920s and 1930s. In effect, despite the advantages of the rack, the two systems have always existed side by side.

When the rack system was employed in thirty-hour clocks it was often placed on the right, as is normal thirty-hour longcase practice. In quality clocks, however, it is virtually always in the left-hand position which became popular early in the eighteenth century and has remained so ever since.

The two late eighteenth-century movements in Plates 23–7 illustrate the simple standard form of rack striking traditional in British clocks until the cam and pin-pallet form of locking (see page 21) was almost universally adopted in the present century. Other forms of locking

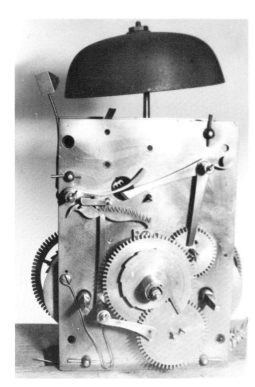
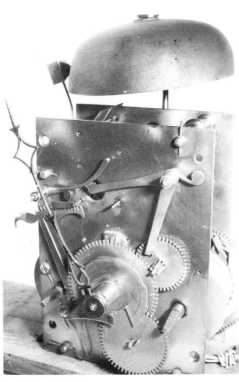

Plate 23   Typical good late eighteenth-century, eight-day longcase movement with rack striking. Note the clearance between warning piece and rackhook, the pallet locking, simple 'safety' spring lifting piece and racktail – and the poorly replaced rack spring
Plate 24   The longcase movement in Plate 23, shown at 'warning'

Plate 25   The longcase movement in Plate 23 has a very common eighteenth-century type of one-piece hammer spring and stop

were, however, developed, partly under European Continental influence (see, for instance, Plate 61). The standard arrangement comprises a cranked lifting and warning piece, raised by the minute wheel and with the warning tip protruding behind the front plate to catch the pin on the warning wheel. Plate 23 shows a gap between the warning piece and the rackhook above it, and this corresponds to the clearance between the link piece and locking arbor already seen on countwheel systems; it ensures that the warning piece is in place before the train is released (Fig 5). Release to warning occurs when the warning piece raises the rackhook and the rack falls, thus freeing the pallet tail which is, when at rest, locked on a pin in the shaped end of the rack. The rack tail falls onto the division of the snail corresponding to the position of the hour hand, and an equivalent number of teeth on the rack thus pass the gathering pallet (Plate 24). The train is still held by the warning piece until this is released as the lifting piece drops off the lifting pin. The train is then freed, and the pallet gathers the teeth until its tail comes to rest again on the rack pin. The shape of the gathering pallet's tail is critical for locking and release, and is either rounded or made with angled faces; it must both rest on the rack pin as the rackhook drops in in front of the first tooth, and come free so that the rack can fall under the influence of its spring. Again, if this spring is too weak, the rack may fail to fall fully at twelve and, if it is too strong, the gathering action may be unreliable. Brass wire has always been found best for this purpose and often the trouble with a botched-up movement is that the brass spring has been replaced with a length of stiff steel-wire jammed into place with a screw instead of anchored by a brass foot with steady pin. Such a replacement can be seen in Plate 23, and the correct spring in Plate 27. As already noted, there is an alternative form of locking more typical of the early period up to about 1720, where the rackhook falls deeply when the rack is gathered and on its arbor is a locking piece acting on a locking-wheel pin between plates (Fig 6d).

The racks and levers of the standard British system are nearly always of steel whilst French racks are often of brass which, acting on a steel pallet – on the lines of brass wheel and steel pinion – would seem to be a better arrangement for parts so subject to wear. The rack and its tail are riveted to either end of a short pipe which runs on a stud in the plate, the rack being normally in the same plane as the cannon pinion whilst the tail has to be further out, in line with hour wheel and snail. This double construction is both a cause of trouble and its remedy, in that the step on the snail struck by the rack tail must correspond precisely to the tooth to be gathered by the gathering pallet, or incorrect counting and jamming result. If the tail is a less than tight friction-fit on the pipe, repeated knocking sends it crooked, and it must

Plate 26 Bracket movement by William Tregent, *c*1780, showing elegant design and tasteful proportions

Plate 27 The Tregent movement (Plate 26) has the type of hammer stop (with a traditional internal spring assembly) adopted when bells were mounted on back plates, as case tops became lower, late in the eighteenth century (see also Plates 73–8)

then be tested round step by step until, by moving the tail relative to the rack, rack tooth and snail step correspond at every hour, when the tail can be riveted, not soldered, more tightly. It should never be necessary, and is inviting disaster, to file rack teeth, which must once have been satisfactory, in order to correct the gathering action. However, if in a desperately worn movement it is necessary, the vertical faces of the teeth should not be touched, only the curved backs. It is possible to stretch a short tooth with gentle punching on the back, but there is a danger of elongating the rack in the process, which will upset all subsequent gathering. In principle, leave racks alone and concentrate on their tails. Again, never try to alter snail steps; they are not subject to wear and the fault must lie elsewhere.

The rack tails in both illustrated movements are of springy hammered brass, with a steel pin to engage the snail. Earlier, and again on many modern clocks, the rack tail is of the same metal as the rack, but with a springy insert holding the pin, the object being to protect the clock should the rack fail to rise at twelve o'clock. The pin is shaped off, as is the vertical face of the snail so that, in this event, the rack tail can be pressed forward and pass over the snail rather than jam against the vertical face. The adjustment of this springing is a matter of common sense, but even with this safeguard the clock is very likely to stop. The damage which would be caused by forcing the hands round with the rack tail butted on the snail is, however, prevented.

The snail itself is screwed to the hour wheel, which has a heavy brass boss to receive the screws. Where, as is usual, the hour hand is fitted to a square, it must of course be fitted so that it points to the hour which the rack tail strikes when it falls; very often the hand is screwed and allows no chance of error here. However, as has been said, whether or not the striking is intended to repeat, the movement may be fitted with a starwheel snail, which can be seen in Plate 51 and in the illustrations of repeating carriage clocks (Plates 35–9). The arrangement in Plate 51 is somewhat unconventional in that, as there is a centre seconds hand, there is no central cannon pinion; but the principle is exactly the same as if the wheel driving the snail were, as is usual in British clocks, the cannon pinion or a second minute wheel. It is immaterial whether the snail revolves clockwise or anti-clockwise, provided it is mounted the appropriate way up. The basis of the drive is a pin in a motion wheel which engages a tooth in the twelve-toothed starwheel which is attached to the snail, and matters are arranged when setting up so that the snail is flicked forward to the next hour only just before the sequence of let-off begins. In such cases, therefore, the positioning of the hour hand is not critical so long as the hour wheel is so positioned that the hand on its square points to an exact hour at let-off – except during the brief period when it is engaged and changing over, the snail

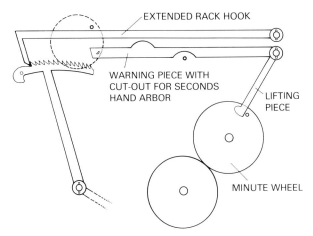

EXTENDED RACK HOOK

WARNING PIECE WITH
CUT-OUT FOR SECONDS
HAND ARBOR

LIFTING
PIECE

MINUTE WHEEL

Fig 18 'Straight-line' rack striking found in some Midlands longcase clocks

can be turned to suit the hour shown. The starwheel is held in place by
a spring and jumper so that, once the pin has taken it past a certain
point, it springs over into the next position.

Standard rack-strike clocks vary somewhat in the shape of their
parts, but there is a characteristic elegance of concave and convex
curves in the rack, rackhook and lifting piece which, while it takes time
to achieve from a supplier's cast blank, should always be sought for in
replacing a missing part of a clock of any quality. There is, however, an
eighteenth-century British provincial layout which is completely plain
and straight (Fig 18) and which seems commonest in the Midlands. The
shapes of rackhook tip and gathering pallet are not only elegant but
vital to performance. We have noted that care is needed in the shape
given to the pallet tail; the pallet itself must present a vertical face
parallel to the faces of the rack teeth and its curve must allow adequate
clearance to the curves in the back of the latter. There is very little
scope for variation in the length of the actual pallet; it should catch a
tooth and pull it slightly beyond the rackhook's position of rest, so that
the rack tooth falls back into engagement with the rackhook. Besides
fitting the back of the tooth, the rackhook tip must be sufficiently
curved at the back so that the tooth can be pulled past it as a ratchet,
and sufficiently straight in the front to hold the gathered tooth
securely. There should be a clear but free fit between the profiles of two
teeth. The precise shape of the top end of the rackhook is largely an
aesthetic matter, but it should be thin enough to allow for bending to
adjust the clearance with the warning piece; bending the warning piece
is never satisfactory since it distorts the whole lifting set-up.

The tail of the lifting piece is commonly bent forward so that, if the
hands should be pushed backwards, the spring-brass is pushed aside by

66

the lifting pin without damage. A good free fit of the lifting piece on its stud is essential; if it is loose, haphazard action of both lifting and warning are inevitable. In the longcase movement (Plate 23) it can be seen that the lifting piece is provided with a projection for manual release; repetition of this simple sort is not uncommon in longcase clocks and examples are known of pulleys by means of which, presumably, a cord could be extended to a bedroom above.

Rack striking's development from internal to external is the reverse of that for countwheel striking where external movements came first. The considerable difficulty of observing the action of the earliest rack mechanisms, and adjusting them, sufficiently explains the emergence of the standard external form. The arrangement with an internal locking wheel already mentioned is transitional, but the movements just examined have all the striking mechanism, save for the warning, on the front plate.

An internal rack mechanism on a thirty-hour movement is illustrated in Plates 28–30 dating perhaps from about 1740 (Fig 19). It will be seen that the lifting piece is in the usual position, but follows countwheel practice in being pinned to a between-plates arbor which holds the warning piece acting internally on the warning wheel and which is provided with a small knop for manual let-off. This arbor also has fixed to it, at the other end, the link piece noted in thirty-hour countwheel striking. However, the link piece connects, with usual clearance, not with a detent and locking arbor, but with an extended projection from a vertical rackhook which is mounted on the arbor adjacent to the top right-hand pillar and sprung with a wire spring screwed into the front plate directly parallel with this pillar. The spring engages a hook above the rackhook arbor, driving the latter up against the rack; in the standard outside mechanism the rackhook works by gravity alone, but it is sprung in French clocks, where again the hooks and racks are usually vertical. The rack itself is very large, being pivoted on an arbor at the extreme left of the movement, with its arm stretching across inside to the right. The rack tail is pinned to an extension of the arbor outside the front plate so that it can fall on the snail from above. Thus the whole rack assembly virtually surrounds the front plate. The second wheel of the striking train carries at the back a gathering pallet built up from its arbor and engaging the rack, as usual, immediately adjacent to the rackhook. This wheel has also a pin for locking, which is effected when this pin meets a stout pin on the bottom of the rack.

Lifting is by means of a cam, of snail form, attached to the cannon pinion; and the hooked shape of the lifting piece prevents damage if the hands are pushed backwards. The action is as follows. The lifting piece raises the warning piece into position in the path of the warning pin

Plate 28 Thirty-hour longcase movement with internal rack striking, c1740. Note single-handed lifting from cannon pinion on the right, and the rack tail crossing the movement to act on the huge snail. (This rack tail is not, in fact, original.) (See also Fig 19)

Plate 29 Side view of the movement in Plate 28, showing pin and stop (on rack) just above greatwheel, gathering pallet raised out of locking wheel's arbor, projection from rackhook which engages with a piece on the lifting-piece arbor, and rudimentary let-off or repeat probably attached to a cord in this house clock; also the curious hammer stop (see Fig 39)

Plate 30 View of the movement in Plate 28 (from above), showing how the rack assembly surrounds the front plate, and depicting also the rackhook and its wire spring alongside the right-hand plate pillar

and continues to rise until the link piece raises the rackhook, thus releasing the rack and allowing the train to run to warning. When the lifting piece falls, the warning is freed and the train runs, whilst the gathering pallet gradually collects up the rack, pulling the rackhook as a ratchet as it does so. Once the rack is fully gathered, the locking-wheel pin meets the pin on the rack and the train stops. At this point the gathering pallet and the rackhook are still some five teeth from the end of the rack which can, in fact, have as many as seventeen. One spare tooth is always necessary in a rack since the rackhook holds, and the pallet cannot gather, the first tooth – sometimes two teeth are idle here. Thus fourteen or fifteen teeth are quite usual.

This movement is beautifully made and finished, with a long delicate pendulum crutch, yet in some respects it seems to be a throw-back in date and system. Even the (repaired) hammer stop and spring arrangement is curious, there being a stout spring engaging a parallel blade on the hammer arbor, acting both to stop and to spring, rather than the one-piece device (Plate 25) which one might expect at the clock's likely eighteenth-century date. It is reliable, but compared with the standard form, for good reasons, obsolete. The action is very difficult to observe and check, the principal parts are inaccessible without complete dismantling, the gathering pallet thrown up from the arbor is prone to wear and difficult to repair, and the spring of the rackhook requires careful adjustment to obtain correct and easy gathering. Some late eight-day movements with internal rackwork and trains on the right are known, but on the whole this internal system of around 1740, with all its manifest inconvenience, seems representative of an early, short-lived, experimental period which was generally over by perhaps 1730, certainly in the main clock-producing centres. It seems to be a country survival of the thirty-hour cottage-type clock of earlier practice (Fig 19). There are few universal rules on striking mechanisms, however. A fine bracket clock by Henry Hindley, with internal rack striking, was recently reviewed in an antiques magazine and dated 1750; and I have seen a provincial quarter-striking internal-rack clock from Portsmouth, in incomplete state, from about 1740.

It is difficult to know where to place the French Comtoise clocks in evolutionary terms. These were made in the Jura, on their own very individual lines and with very little change, for some two hundred years from about 1690 onwards. Though they started with one hand and developed to two hands, and changed from verge to anchor escapement in the mid-nineteenth century, they seem always to have had their own special form of rack striking – a vertical rack with a looped wire tail illustrated in the verge movement of Plates 31–2, which dates from about 1850.

It will be seen that the basic structure is posted, like that of British

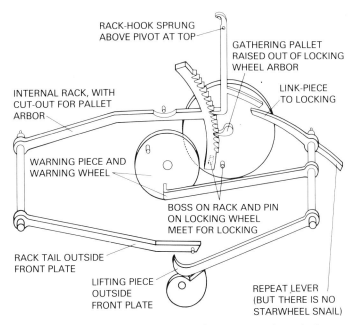

RACK-HOOK SPRUNG
ABOVE PIVOT AT TOP

GATHERING PALLET
RAISED OUT OF LOCKING
WHEEL ARBOR

INTERNAL RACK, WITH
CUT-OUT FOR PALLET
ARBOR

LINK-PIECE
TO LOCKING

WARNING PIECE AND
WARNING WHEEL

BOSS ON RACK AND PIN
ON LOCKING WHEEL
MEET FOR LOCKING

RACK TAIL OUTSIDE
FRONT PLATE

LIFTING PIECE
OUTSIDE
FRONT PLATE

REPEAT LEVER
(BUT THERE IS NO
STARWHEEL SNAIL)

Fig 19   Thirty-hour internal-rack striking (section through plate)

'bird-cage' movements, but on a much larger scale. These frames were provided with iron doors and back, and a pressed-brass front enclosing the enamel dial, so that they were totally enclosed without a case; they were in fact 'exported' and the wooden cases made elsewhere for the most part. The clocks have apparently massive pressed-brass pendulums, sometimes with rocking devices in them; in fact the metal is very thin, the pendulums being weighted with sheet metal at the bottom and made so that they can be folded or taken apart for transport. They were suspended by silk (as here) until much the same time as the change from verge to anchor, and they hang in front of the movement, with a circular addition to clear the hand pipes. Originally the striking levers were held in place by springs, but from about 1800 these were replaced by brass counterpoises – one to the cranked lifting piece and one to the rackhook arm in the present movement. Later, for symmetry and crisper action, another weight was added to the lifting piece; here its place is taken by a pierced tag for a repeat cord. The striking trains are always on the right of the movements.

At the heart of the oddity of Comtoises is the lifting piece, the lifting end of which is forked and acts on broken hoops projecting from the back of the cannon pinion. It is suspended from an arbor at the top of the right-hand strip plate and acts by gravity. The purpose of the fork, universal with these clocks, is to give a double strike at the hour. When the tang on the wheel falls inwards, the clock strikes and the second

70

tang lands on the rim; three minutes later, it also falls and the clock again strikes to reassure the doubtful or innumerate owner. At the half, there is no room for the second tang to fall and only a single blow is struck, the rack not being released.

The lifting piece is on the same arbor as a right-angled lever with a hinged detent at the rear of the movement (Fig 20). When the lifting piece falls, the weight of this assembly knocks out both the locking and the rackhook – there is no warning in these clocks. The second arbor at the top thus has at its front the rackhook and at its back the locking piece, which engages with a pin on the third, locking and gathering, wheel. The tip of the locking piece has a notch cut out, the face of which performs the release for the second strike.

All these levers are steel. The rack is a 7.5cm (3in) strip of brass with ratchet teeth, and it ends in steel pins sliding in small brass brackets riveted to the plate. Freedom for the rack is clearly essential; it has very little weight and many a repairman must have been tempted to add a small spring or a drop of oil. The oil will make matters worse in

Plate 31   Verge Comtoise movement, *c*1850, with thread pendulum suspension. Sheet metal to back and left-hand side removed to show movement, which is normally totally enclosed in a box with the dial plate forming the fourth side. Note typical circle of pendulum round centre arbor, and indirect crutch linkage to the left. For layout of levers see Fig 20

Plate 32   The Comtoise movement (Plate 31) vertical hammer rod and wirespring, starwheel hammer action, and locking piece (on the arbor of the vertical rackhook to the front)

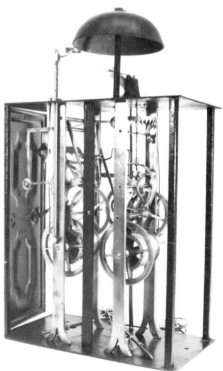

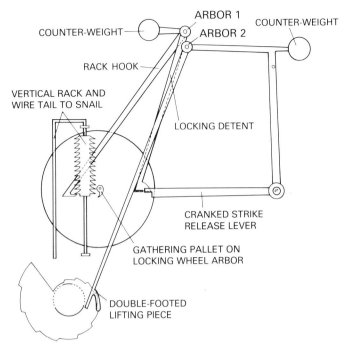

ARBOR 1
ARBOR 2
COUNTER-WEIGHT
COUNTER-WEIGHT
RACK HOOK
VERTICAL RACK AND
WIRE TAIL TO SNAIL
LOCKING DETENT
CRANKED STRIKE
RELEASE LEVER
GATHERING PALLET ON
LOCKING WHEEL ARBOR
DOUBLE-FOOTED
LIFTING PIECE

Fig 20   The Comtoise basic striking mechanism with two arbors (from front)

course of time; one must instead ensure that the pins are properly straight and aligned, and burnish them and the bracket holes with a round broach or a large needle. To this strange rack is attached a twisted steel wire which loops up and over and then descends vertically, through a brass guide extending from the plate to the centre, to above the snail, which is always skeletonised and attached to the hour wheel. Thus although the action of rack tail and rack is much more direct than in the oblique British system, there is the same need to ensure that rack teeth and snail position are in exact agreement. This is done by bending the wire rack tail as the strike is tested round.

There remains the gathering mechanism, and this is typically European Continental in that the rack is double-sided – one side for the rackhook, the other for the gathering pallet. The latter is directly behind the strip plate, being a single leaf of the locking-wheel pinion, with the rest broken away and flattened. Thus one of the difficulties in setting up rack systems – setting the separate gathering pallet in the best position relative to the rack and locking – is avoided, since the locking pin and the pallet are on the same arbor and fixed. There are in fact few difficulties on the striking side of these movements save for the assembly of the free movement of the rack already referred to and the setting of the pinwheel (always a starwheel) free of the hammer. The

72

commonest item for substantial repair is the gathering pallet, which wears. Either a fresh nib can be silver-soldered into position, or a pin can be drilled into the pinion leaf and bent up to unworn height – that of the pinion leaf. Hammer springs are powerful and hammer holes usually require bushing.

Assembly should be carried out with the movement lying on its front and with the train locked; although the front plates, cannon pinion and bracket to take the lifting piece must be fitted first of all with the movement facing upwards. Hammer design varies, but the vertical rotating design illustrated is the oldest and commonest.

It is simplest to insert the hammer rod, turn the starwheel until it is free of the sloped lug against which it presses to turn the hammer rod, and then fasten the assembly temporarily with Sellotape whilst the other parts are added. Place the lifting piece in position with the inner tang just about to fall off the larger section of the cannon-pinion hoop, and the hinged arm resting on the locking-wheel arbor. The rackhook has to be juggled into position so that its pallet is below the inner set of rack teeth and the locking piece about a millimetre from the hinged piece, with the locking pin up against it. A small point to note is that the long vertical hammer springs should be treated with care, as they are hardened and very brittle.

The origins of this mechanism are uncertain. The rack appeared in France in connection with repeating systems at about the same time as it appeared in Britain, but the relationship of the different forms has not been traced. Comtoises with quarter chiming, and quarter striking, exist, though they are rare; the hour strike is let off, as in the British system, by the deep fall of the quarter rack at the fourth quarter. A few clocks have been found with single-sided racks, but otherwise their mechanism, including the wire rack tail, is similar. The Comtoise is a reliable system if properly set up, and demonstrates what the British seem seldom to have realised, namely that the extra expense and work of the warning system is not essential to rack striking except where the hour is let off by chiming. The appearance of these clocks, however, is not always to modern taste, being mostly very large and florid with curved cases.

This Jura area of France was also a major centre for the production of carriage-clock movements, often distinguished by having all their striking mechanism visible on the back plate. But these, so far as is known, never strike twice at the hour; double striking was evidently a feature of the main Comtoise house clock.

The extent to which the Comtoise is out of line with French rack movements generally will be very obvious from comparison with the clock of about 1860 in Plate 33, which represents a standard movement of a type which had existed for over a hundred years previously.

73

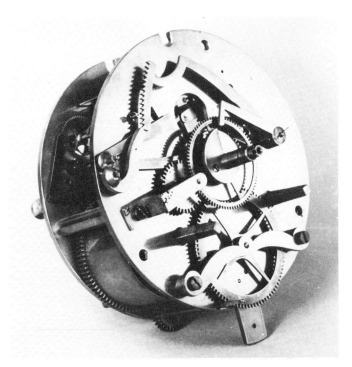

Plate 33 Typical French rack-striking movement, *c*1860. Note connection of rackhook to locking piece between plates (pin locking). The lifting piece raises the rackhook by a pin (not visible) on the rackhook. The warning is by a blade moving in the slot just to the right of the rack

Moreover, whilst the workings are basically similar, the whole conception is different from that of standard British rack movements and indicates how, partly for economic reasons, the Continental industry overtook that of Britain and influenced its later models. The French movement is as substantial in its own way as the British, but its weight is perhaps about a third, partly because there is a going barrel instead of the fusee which was usual in Britain, and partly because all the parts and plates are so much smaller and lighter. Needless to say, the power input is correspondingly less and the fitting of parts on studs and in pivot holes more critical. The lifting and warning piece is on the left and works on the cannon pinion; the rack is mounted to the right and has a vertical fall, assisted usually by a small leaf-spring. There is no safety-spring in the rack tail, so that if these clocks fail to gather at twelve, the clock stops and the motion wheels may have to be dismantled to remove the rack and reinstate it; if the hands, which are generally of fine blued steel, are forced on, damage occurs. The snail in the non-repeating movements is screwed to the hour wheel and is continuous, having no steps for the hours, and this can prolong the setting-up process. The wheels are finely crossed out with light narrow rims and the pivots should be highly polished; the striking train is

74

almost noiseless when running, in contrast with the considerable gear noise from even a good British movement.

Two distinctive features centre on the S-shaped rackhook. First, this is, as in the British arrangement, raised by the warning piece, but by means of a pin projecting from the rear of the hook which encounters the warning piece as the lifting piece rises (Fig 21). (The warning, as in the British system, is by a blade passing through a slot in the front plate and meeting the warning pin on the wheel.) This pin may require adjustment to provide the normal clearance which ensures that the warning piece is in place before the train is unlocked. Locking is controlled by the rackhook, which is on the squared extension of an arbor carrying a locking piece which engages with the locking-wheel pin. It holds the pin only when the rackhook is tucked beneath the fully gathered rack; when the train is in motion, the hook rides out on the rack teeth and the locking piece is out of range of the locking pin. There is a quite common alternative locking set-up in which the gathering pallet has a 'British' tail which locks onto a pin on top of the rackhook. Much more rarely, the rackhook is taken through the plate where its extension catches the locking-wheel pin. It will be noted that the locking arrangement of the present movement is virtually identical to that of the countwheel movement in Plates 15–16 and the lifting/ warning piece is similar; the train cannot be locked either when the countwheel detent is on a high section of the countwheel or when the rackhook is on the high (toothed) section of the rack which has fifteen teeth, the pallet not engaging the bottom two.

Three important parts – the rack, the rackhook and the lifting piece – are mounted on fine studs screwed into the front plate. Should these

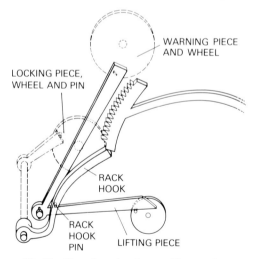

Fig 21   French rack release with warning

75

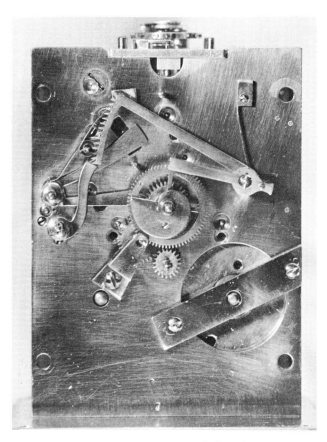

Plate 34 French simple rack-striking carriage clock with warning, c1890, directly comparable with movement shown in Plate 33

be loose or bent, or should the fit of the parts be sloppy, the action will immediately be unreliable, and this, along with simple bending of the brass rack and the tendency of the locking wheel (pallet) hole to enlarge causing inaccurate gathering, is the most common area to need attention. The other point is that the pipe of the cannon pinion which takes the minute hand is squared, and of course the clock should strike when the hand is exactly at the half or hour. One sometimes finds that the lifting pins have been bent with this in view, but there is so little space that this can cause unreliable action. The minute hands are almost always riveted to a brass squared collet and the proper course is to turn this, after loosening with heat if necessary, until the hand points correctly, when the collet is riveted more tightly. The fly is very light and mounted in an eccentric hole, where it can be adjusted for smoothest and quietest running. As a rule these small pinions do not wear, the power at this end being so small. The wheels at the foot of the movement are to operate a separate Brocot calendar unit; these calendars are normally driven by the striking train.

Plate 35 Early twentieth-century French repeating carriage clock with rack striking, flirt release and no warning (see also Fig 4). Note, below the rackhook, the lever operated by the minute wheel which prevents the rack from falling at half hours; also the starwheel snail always used in these repeating movements; also the left-hand arm of the repeat piece which intercepts the fly. The locking piece (between plates) is typically cut away to clear the hammer arbor. There is a simple pin stop for the hammer

Carriage clocks embody the highest level of finish and precision. That in Plate 34, dating from about 1890, has the simplest form of hour and half-hour striking, essentially the same as in the movement just considered, although it will be seen that there is a spring-steel rack-tail safety piece, and that the warning piece – as with the French countwheel movement – is sprung downwards rather than relying on gravity. This movement sounds on a coiled-tape gong, and it will be noted that its ratio is correspondingly very much higher than that of the countwheel movement (see page 39). Again there are fifteen teeth on the rack, an eccentric hole for the fly, and the locking is in fact precisely as in Plate 33.

Such carriage clocks are sometimes 'converted' to repeat to increase their value, but in fact neither the trains, the power, the arrangement of the snail, nor the presence of warning make them suitable for repeat striking. For this a special type of movement was developed, of which Plate 35 is a late, twentieth-century example in standard form. The essence of repeating carriage clocks is their flirt release, in which there is no warning. The flirt is a hinged arm (Fig 4), pivoted on a stud just to the right of the repeat pusher, running down to a point beneath the cannon pinion where it is pushed aside by the lifting pins, then rising to end in the forked guide at the top left of the movement. The hinge is just visible at the apex through the hour wheel. Whilst the rack corresponds roughly in shape to those already seen on French movements, the rackhook is now more vertical and continues up beyond the foot of the rack. To it, behind its upper tip, is fixed a pin; the flirt, once drawn back and then freed by the lifting pin, flies across the clock, and a notch at its end (here obscured by the guide fork) catches this pin and pushes the rackhook aside, thus freeing the train – for the locking piece is, as before, on the rackhook's arbor. Of course the rackhook cannot be left held out and must be freed to spring back against the rack. This is done by having a double pallet, the bigger rear section of which raises the flirt and releases the rackhook.

The snail, desirably for a repeating clock, is mounted on a starwheel with jumper, advancing only just before the full hour. The half-hour striking of the previous rack movement was in fact produced by a more central lifting pin, which caused the rackhook to move far enough to release the train but not to release the rack – a nice adjustment which often goes wrong. In the present movement, notice the minute wheel on its prominent cock below the hour wheel. This bears underneath a pin which every hour moves a cranked lever in and out below the foot of the rack; it has to be set up so that it is beneath the rack at the half hour and then, as the rack cannot fall, equally spaced lifting pins can be employed.

The repetition lever is a short pivoted arm with a long branch downwards. This terminates in a pin which knocks out the rackhook and frees the train when the button is pressed; it also pushes aside the half-hour lever with a pin. However, the short arm, at the other end from the pushpiece, ends in a long pin which goes into the movement and intercepts the fly for as long as the button is depressed. Thus there is no chance, as there is with some cruder repeating systems, that the train will start to run and gather before the rack has fully fallen causing a short number of blows to be sounded.

Provided that it is correctly set up, the principal difficulty with this type of movement is often the rackhook. This is sprung and has a blunt point which must be at the correct angle – not far off parallel – to the horizontal straight face of the teeth if it is to hold them at rest and then let them pass, as a pawl and ratchet, when the pallet is raising a tooth. The difficulty is more often imagined than real, and is brought about by

78

filing the rackhook faces in an attempt to correct faulty gathering when the fault lies elsewhere, usually in a bent rack. Although these racks have been said to fall vertically, they are, of course, pivoted and describe the path of a large circle. The gathering pallet is in the radius of this circle, directly opposite the rack's stud. If the rack itself is at all bent, one way or the other, on its arm, there is very little hope of correct gathering because the depth of engagement of gathering pallet and teeth will vary. Before vandalising a rackhook, therefore, or filing away at rack teeth – none of which is likely to be seriously amiss and must once have been correct – it is wise to take off the rack, place it on a piece of paper and, with its hole as centre, draw a circle round it, adjusting the angle of rack and arm until the rack teeth all touch the circumference.

A further complication in carriage clocks is quarter repeating and grande sonnerie (Plates 36–9). Here, beneath the rather forbidding appearance of the grande sonnerie, can be seen the outlines of the simpler hour repeater of Plate 35. They work on the same principle, with the addition of quarter mechanism and often with refinements which were the personal choice of the owner.

The quarter mechanism consists of a quarter rack on the same stud, with similar radius and the same sized teeth, and therefore unfortunately visible only as a shadow in these photographs, where it has the lower of the two rack arms. For this rack there is a separate part at the rear of the rackhook, a second pallet (third, if you count that for raising the flirt) and a separate snail. The pallets are typically double to reduce the revolutions required of the train. The quarter snail is on the cannon-pinion arbor and the quarter-rack tail protrudes from the middle of its rack arm. Usually, this snail has a complication known as a 'surprise piece', which consists of a high segment fastened freely over the snail. Its effect is to prevent the quarter rack from falling on the hour. It is moved forward by the starwheel hour-snail changeover at the hour. Neither quarter snail nor surprise piece is visible in the illustrations.

The crux of the quarter-striking action is the double, shaped rackhook, which projects less at the rear, for the quarters, than in front for the hours. Both parts of the rack could be gathered together, being identical apart from length; but when the hour rack falls and is gathered, the rackhook is operating too far out to hold the quarter rack. Only when the hour rack is fully gathered, and the rackhook has moved in to hold it up, can the shorter part of the rackhook operate on the quarter rack so that it, too, can be gathered. Thus, if the option of full grande sonnerie has been set, at each quarter are sounded the preceding hour dictated by the starwheel snail and the latest quarter dictated by the quarter snail, in that order.

Plate 36   French grande-sonnerie repeating carriage clock, *c*1885. Shown at rest ready to strike grande sonnerie – the long vertical setting lever is pushing aside the cranked minute-wheel lever which otherwise prevents hours being struck at the quarters. Quarter rack is behind hour rack, and quarter snail is on central arbor. The rackhook arrests both racks

Plate 37   The grande-sonnerie clock (Plate 36) shown with racks fallen for striking of hour and quarter. Note the top hook silencing higher bell for preliminary striking of the hour

It will be remembered that in the hour-repeating carriage movement there was a cranked lever, worked by the minute wheel, which prevented the hour rack from falling at the half hours. In the grande sonnerie clock this lever is found again and is sprung to prevent the hour rack falling at any quarter; we then have plain quarters and hours (petite sonnerie). But it can be overridden manually by a branch of the vertical control lever which passes down into the base, and that is the situation in all these photographs – the silencing lever is set aside to let the hour rack fall so that the clock strikes the hour also at every quarter. The lever has also a third position in which it prevents the flirt from catching the rackhook, ie the clock is silenced. The repeating mechanism is precisely as in the hour striker, except that it cannot overrule the control lever and its governing of the fall of the hour rack. Therefore if the lever is set to 'grande sonnerie', the clock, very desirably, will repeat grande sonnerie.

Quarters are, of course, struck ting-tang on two – more rarely up to

Plate 38   The grande-sonnerie clock (Plate 36) shown between striking of hour and quarter. Both hooks impede the hammer as the hour rack is raised to its highest point, and a silent 'dummy blow' is sounded

Plate 39   The grande-sonnerie clock (Plate 36). Pressing the repeat button knocks out the rackhook so that both racks fall (the quarter rack here is on the first, highest section of snail) and the previous hour and quarter are sounded. The repeat button overrides the vertical lever so that the hour is sounded at 'repeat' even if the clock is set to sound quarters without hours

at least four – bells or gongs, whilst hours are struck only on the deeper one of the two. This is generally brought about by arranging for the high-noted bell's hammer tail to be 'pumped' out of the way of the pinwheel at the hours. In French carriage clocks, however, it is effected by the axe-shaped horizontal steel lever and the two little steel hooks on the left of the movement. These hooks are attached to the hammer arbors, and a hammer can strike only when its hook is in the notch of the steel lever or below the lever altogether, as is the position at rest. The movement of the steel lever is controlled by a pin at the foot of the hour rack. When the racks fall for quarter striking, this lever falls since it is no longer supported by the hour rack pin, and presents a blank face to the upper hook so that the hours may be struck first (grande sonnerie) on the lower bell only. As the hour rack is gathered to its highest point – the quarter rack as yet not catching on the rackhook and so being ungathered – it raises the lever to such an extent that both hooks and hammers are obstructed and there occurs a 'dummy

81

blow'. Thereafter, the hour rack falls onto the hook and the lever allows both hooks and hammers freedom, as in the position of rest, as the quarter rack is gathered up and the quarters are struck. At the hour, it is unusual for four quarters to be struck; the quarter rack cannot fall and, with the release of the hour rack, the lever falls to the position where only the lower note can be sounded.

These carriage clocks are valuable and fragile, the chief dangers being bent parts and bad fits to the many studs. The racks should be tested both for flatness and circumference as already described (page 79). On no account tamper with the rackhook faces. Whilst the layout of most such clocks is standard, it is as well to proceed cautiously with sketches and piercing labelled holes in a box, for screws and studs, when dismantling. The screwed studs are not interchangeable, and if confusion arises it may well involve removing half of what you have already put together and set up. The vital flirt and its stud, for example, are almost totally inaccessible for alteration once the movement has been assembled, and the wrong stud here means going back to the beginning. As the setting of snail, starwheel and cranked rack lever may well not be correct first time – and the setting of hammer tails 'off the pins' is no mean feat in these many-pinned pinwheels – there is very little point in mounting the repeat assembly until full tests have been run. With regard to the value of grande

Plate 40   Quality ting-tang quarter-striking rack movement, c1900, shown striking a quarter past one. Note the lever working on the minute wheel is positioned downwards, freeing the upper bell's hammer
Plate 41   Side view of the ting-tang movement (Plate 40), showing lever and pin for acting on the lower (higher-pitched) bell hammer, and pin locking by the rackhook in European fashion

sonnerie carriage clocks and the frauds that are perpetrated as a result, the reader can do no better than refer to Allix and Bonnert's *Carriage Clocks*, which also contains most useful discussion of grande sonnerie trains and other characteristics.

Plates 40–5 show other means of obtaining quarter striking or 'three by two' trains. The movement in Plates 40–1 is a high-quality ting-tang movement, massive and typically British, but in fact more probably German. German and French ting-tang movements were very popular around the end of the nineteenth century and, even at this time, the higher quality British movements were still often made with fusees, whereas here we have massive going barrels. European Continental influence is also seen in the layout of rack, rackhook and lifting piece, which resemble those in eg Plate 33, as does the locking.

The principle is the very ancient one of 'pumping' a hammer out of the way of the pinwheel when it is not required. In a ting-tang clock, the high-noted hammer is moved so that the hours are struck on the low gong alone; here this is done by the clearly visible pivoted lever working on the minute wheel; its tip protrudes through the plate and raises the high-noted hammer tail out of the path of the pins. An alternative method was to use a pinwheel with teeth on either side and to pump one hammer arbor so that only the other hammer engaged. This system existed even in the seventeenth century (Fig 22) and might be used for quarter striking with a 150-division countwheel.

In the method shown in Plates 40–1, the lever must be set up to act just before the hour, and by the first quarter it will have been released again. In conjunction with this device is the use of three extra teeth on the rack for striking the quarters, corresponding to three lifting pins on the cannon pinion set at graduated distances from the centre. The rackhook releases the teeth as dictated by these pins and also the deliberately curved upper edge of the lifting piece where it engages the pin on the rackhook, and, as the teeth are also graduated in size, the result, if wear has not taken its toll of the teeth, is correct ting-tang quarter striking at three quarters and no quarters at the hour. The snail is provided with a cut-out in its highest section, and this has to be carefully positioned when setting up so that the clock strikes one on the short highest section – having correctly sounded twelve on the lowest – and the rack is free to fall up to three teeth during the rest of the hour of one. Such movements will be found much simpler to service and set up than the corresponding carriage-clock quarter strikers; but whether they would be so if they were the same size is another matter.

The other movement (Plates 42–5) illustrates just how far quarter striking can be taken without becoming quarter chiming. There are limitations, but this clock was clearly designed to come as near as possible to full quarter chiming with only two fusee trains. It dates,

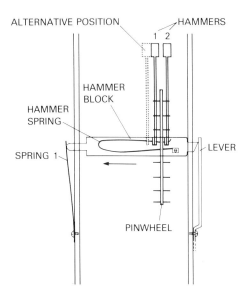

ALTERNATIVE POSITION

HAMMERS

1 2

HAMMER
BLOCK

HAMMER
SPRING

SPRING 1

LEVER

PINWHEEL

Fig 22 Eighteenth-century 'pumping out' of one bell. When the lever depresses the arbor of the hammer block, it moves to the left, taking hammer 1 out of engagement with the pinwheel to the alternative position. Spring 1 keeps both bells in operation for quarters. The hammer spring works on the hammer backs in the block. Lifting pins may be straight through the wheel as here (using different lengths of hammer tail), or staggered on alternate sides of the pinwheel.

probably, from about 1900. The style of the hands puts it in the mid-nineteenth century; but other factors, such as its Westminster chimes, the general lightness of construction for a fusee movement and the use of nuts to secure the steel pillars, suggest a later date. It may be by the Winterhalder family, German pioneers of mass-production in the nineteenth century, who produced chiming clocks, sometimes of this type, of above average quality.

The principle in this movement is to use one rack with two tails and two snails. The main hourwork, though somewhat unusual in the relative position of the parts and in being on the right-hand side, as with most full chiming systems, is like the substantial British flirt-rack movement customarily adopted in chiming clocks of some lavishness at this time. The flirt, fitted with a spring protection against reversal, lifts from four equally spaced lifting pins on the minute wheel to the left. Locking is by a large triangular pallet tail onto the pin in the rack. This has another purpose. The rackhook, when flung up by the flirt, catches on the pin of a sprung lever (here concealed) to ensure that the rack is left free to fall. The large pallet tail, similar to the rear part of a French repeating gathering pallet, disengages this catch on its first gathering revolution, so that the rack falls and is free to gather. The snail, apart from a cut-out at one, is normal.

84

This, then, is how the clock strikes the hours. But it also has a pin barrel set out for Westminster chimes, a large fly of the type used for chimes, and a second, smaller rack tail which works on another minute wheel directly below the hour wheel. The design of this second wheel is the crucial factor for the quarters for it has, raised from it, an internal snail with a gap in it, inside which the second rack tail operates. This second snail and rack tail are so placed that they engage before the hour-rack tail can strike the hour snail, except in the hour of one, which is provided for as usual by the cut-out in the hour snail. Therefore, at the first quarter the flirt releases the rackhook and the quarter-rack tail falls and strikes the highest section of the quarter snail; and so on, through the other quarters, until the hour. At the hour, the quarter-rack tail passes out of the internal snail through the slot, and the hour rack tail is free to fall onto the hour rack. Quarters are not struck at the hour.

It remains to see how this ingenious system functions with Westminster chimes rather than ting-tangs – apart from noting the enormous spring and barrel required to power both striking and quarter striking. But function it does – inevitably with all the confusion of sequence to which Grimthorpe referred in connection with repetition work applied to chiming (page 29). The pin-barrel is double, carrying pins both for the hours and for the quarters. This was not uncommon in the eighteenth century when the hour pins were usually taken out of the end of the barrel, but here they are simply an additional series of five pins. Thus we have a barrel containing five sequences of the Westminster chimes, each having four pins and beginning parallel to an hour pin. This barrel is mounted on a sprung arbor and pumped backwards, by a lever operated from the cannon pinion, for the hours; it is only then that, if correctly set up, the hour pins come into line with the hour hammer and gong – at all other times the barrel is aligned to the four quarter-hour hammers.

How, it may be asked, since the revolutions performed by the barrel must vary according to the number of blows struck, is it arranged that, for example, the first quarter struck is the true first sequence of the Westminster chimes. The simple answer is that it isn't; the first quarter, and any other quarter, varies unpredictably – or at least according to a very complex sequence – depending on the hour. Thus, although the order of the chimes on the barrel is always followed, and counted according to the quarter snail, sounding may start at the beginning of any sequence. This could have been tolerated only when the chimes were familiar, or perhaps from a foreign maker to whom they meant less.

Besides this, nonetheless, the Westminster rod chimers of the 1920s and 1930s seem pale reflections. Compared with them, if not with all its

Plate 42   Full quarter-striking movement, c1900. Note flirt release of the rackhook and use of a second rack tail and minute wheel with 'internal' snail to count quarters. Pallet-tail locking and flirt design are English in style, but the movement is probably German
Plate 43   The full quarter-striking movement (Plate 42) striking three quarters. Note small rack tail resting on internal snail of lower minute wheel

predecessors, the movement is massive, with fine pillars, the wheels finely cut and crossed. There is a solid yet simple dome-topped mahogany case, and the movement is mounted with traditional brackets. The arrangement of hammer springs, with its upright bracket, is in the tradition of chiming British clocks, not of the mass-produced lighter varieties. The gongs of coiled blued steel are substantial and the sounding very resonant. The steel rack, rackhook and flirt, even if foreign, are just such as would be found in luxury traditional rack-chiming British clocks, both bracket and longcase, of the same period rather than in the many cheap brass movements which were beginning to flood the market even as Grimthorpe declared that 'the ornamental English clock trade has ceased to exist, having been entirely driven out [by American and European Continental clocks]' and 'his' chimes – borrowed in fact from Great St Mary's, Cambridge – started on their extraordinary career as favourites for almost a century. Incidentally, the purpose of the 'repeat' cord in this clock escapes me; the movement will not repeat hours save on the hour, and to correct the sequence of quarters is pointless when it will be wrong again in an hour's time.

Plate 44   The full quarter-striking movement (Plate 42) striking one o'clock. Note small rack tail opposite gap in internal snail (through which it passes at later hours)
Plate 45   Note huge strike mainspring barrel, adjustable fly and traditional chime barrel and hammer assembly of the full quarter-striking movement (Plate 42)

The German firm of Uhrenfabrikation in Lenzkirch made a rather less substantial ting-tang version of this mechanism at the turn of the century. Again, the rack is provided with two tails – one acting on a quarter snail attached to the cannon pinion which also does the lifting, and the other in the usual way acting on the hour snail attached to the hour wheel. The fourth quarter of the quarter snail is cut right back so that here the quarter-rack tail does not prevent the hour-rack tail from falling onto the hour snail. There is the usual cut-out in the hour snail to permit sounding of the quarters after one o'clock, and the minute wheel has a pin which operates a lever to lift the high-noted hammer tail out of action when the hour is being struck. Pallet locking is employed. Whilst less massive than the model in Plate 42, these Lenzkirch movements are of good quality and well finished, and they tend to be found in massive 'presentation' cases from about 1895 onwards. The dominance of foreign movements in quality clocks for this purpose illustrates the state of market forces at the time – a traditionally made British movement of comparable performance would undoubtedly have been very much more expensive.

The British industry did recover, though hardly to its former strength and independence of foreign competition. Modern movements

Plate 46 Modern rack-striking movement, showing integral rack and tail, pallet and warning wheel, locking, and straight ratchet-teeth. The rackhook is operated by the pallet cam, not the rack teeth

have a European, if not international, style, although that in Plates 46–7 is, in fact, British – a very typical modern rack-striking movement of no special pretensions. Various significant developments, essentially the outcome of economic forces, will be noticed. For instance, following earlier Continental fashion, the rack is set over to the right. Its arm is at the base of the rack so that there is no separate rack tail as such, its function being performed by a springy steel nib like the pin of a conventional rack tail half-way along the rack's arm. It is thus completely unadjustable but, equally, incapable of slipping. The snail is of thin metal and is riveted, not screwed, to the hour wheel. The hammer is pressed into its arbor and again its head is riveted rather than screwed to the shank. The lifting piece is pivoted to the right, and can be lifted out of the way by the silencing lever which projects to the right.

However, the most interesting changes to be seen, and very general in modern clocks both striking and chiming, are concerned with warning, gathering and locking. As when the rackhook was attached to an internal locking piece, so here the rackhook both holds the rack and brings about locking. This latter occurs when the face of the cam gathering pallet butts onto the projection half-way down the rackhook,

Plate 47   Detail of locking and rackhook on the movement in Plate 46

at which point the top of the hook also catches the warning pin. Once the pallet has passed this point, on the release of the rackhook by the lifting piece, the train runs into warning, a backward extension of the lifting piece projecting into the movement. When the warning piece and rackhook move, the rack also has fallen and the rackhook cannot move far enough in to lock the gathering pallet. This pallet, the cam, has on it a projecting pin which does the actual gathering, and its shape is such that it lets the rackhook in to catch a tooth whenever the pin is out of contact with the rack, so that the rackhook no longer acts as a pawl and the rack teeth need not be ratchet-shaped (Fig 37). Once again, a degree of adjustment, that of the link between lifting, locking and warning, has gone and there is no facility for long-term correction, though the warning piece can be bent up or down. On the other hand, the number of studs – simply riveted to the skeleton plates – is reduced, resulting in a very rigid and reliable action to which repair, apart from replacing the odd bent pin or part, is seldom necessary provided all is kept clean and oiled. Whilst the train must be set up with the hammer tail clear of its starwheel (possibly with a set-screw) and the warning pin to catch the top locking when at rest, there are no further complications since the position of the pressed-on pallet is adjustable after assembly for warning and locking.

Finally, Plates 48–50 show how similar principles have been adopted

Plate 48 Contemporary cuckoo clock with circular rack and adjustable rack tail. The rackhook action is of a ratchet type, the cam locking on the hook pin but not moving the hook in and out. Combined lifting and warning piece visible between rack and snail. Locking is also on the warning wheel

Plate 49 The contemporary cuckoo clock (Plate 48), seen at warning

Plate 50 Action of the bird of the contemporary cuckoo clock (Plate 48), by crank and cam on pallet-wheel arbor

for contemporary cuckoo clocks. This movement is still in production and a very different matter from the countwheel example in Plate 17, which is at least ninety years earlier. The movement employs a rack and rackhook shaped to occupy minimum space by curling round the hour wheel and snail; the one-piece rack for this was patented in 1955. Although in one piece, there is a distinct rack tail which is, in fact, adjustable by bending, should the snail positions not correspond precisely to the rack teeth. All is reversed in position, but the principles are much as in the previous movement. The gathering pallet locks, this time with a hook round a boss on the rackhook; and at the same time an extension of the latter catches the warning pin at the top. Gathering is by a pin pallet, but in this movement the rackhook works as a ratchet during this process. Warning is on the warning-wheel pin when it has turned downwards, where it is intercepted by a rearward extension of the lifting piece into the movement.

The arrangement of the connecting links is the same as in the older clock (see page 55) but the tails are now on arbors extended outside the back plate and driven by an external starwheel, whose position is adjustable. This greatly simplifies the setting up of these clocks, as do the skeletonised brass plates which admit light into the movement. The bird is now plastic with fixed wings, and the earlier two doors are now only one, though still opened by pressure from behind by the bird. The latter operates in a different manner. A cranked lever rests on a notched cam on the pallet-wheel arbor and this lever, when raised by the revolving cam, presses on a pallet on the revolving vertical rod attached to the bird and causes it to turn, with the bird opening the door. However, there are many different and patented devices for bird operation. Their working will be apparent from observation.

These clocks seem to give a good deal of trouble, but it is usually superficial – someone has managed to get the chain off the sprockets, the links have come undone or been bent and so on. There are, however, two recurrent minor faults. The first is that the pin pallet lies extremely close to the rack and, if very slightly bent, comes into the path of the latter so that it will not strike a full count. The remedy is simply to bend the pin. The second difficulty is that the very light rack, unlike the rackhook, is not sprung, but relies on gravity for its fall. After a time it seizes up on its stud and has to be freed, broached out very slightly, and oiled. How such a clock will wear is a very different matter. It seems built almost on a disposable principle, so there is little spare metal for adjustment or repair. There is no reason why the holes should wear any better than the bushes of the old wooden-plated Black Forest clock, and indeed other similar models show that they do not wear as well. There is, possibly, a saving grace in solid pinions rather than lanterns; but whilst lanterns wear they continue to run when worn, whereas worn solid pinions give trouble and require replacement rather than repair.

These modern movements are clearly mass-produced for fairly transient decoration and enjoyment. We have come a very long way from the solid tradition of the British rack movement built to last and to sound the hour in a compelling and impressive manner. The evolution of mass-produced chimework in the 1920s and 1930s was a similar development, although then there were less reliable alternative time-tellers in the form of radios and quartz movements. We shall look at chimework next.

5

COUNTWHEEL CHIMING

We have now to consider clocks with a third train, for chiming. As has been seen, almost any combination can be found, with the exception, so far as I know, of rack chiming with countwheel striking. The earliest chiming clocks, which one is unlikely to have to repair for they are very rare and valuable, naturally used countwheels for both hours and quarters. Indeed, there exist some 'bird-cage' and lantern movements of this type. Where the three trains are arranged fore and aft, the chiming is usually in the middle. Where the trains are side by side as in full countwheel systems, the striking tends to retain its usual position – on the right in thirty-hour clocks, generally on the left in eight-day clocks – and the chime train tends to be on the other side. Whatever the positioning, the almost invariable arrangement is for the going train to let off the quarter train and for the quarter train to let off the strike; it is rare for the hours to be let off by the going train, for this creates problems of synchronisation. A chiming countwheel lets off an hour countwheel by a system generally of two double levers and a pin on the chiming wheel; the latter raises the first lever, causing the second lever to release the locking on the hour countwheel, the warning having been set up as the levers started to move.

There seems to be little information on the relative frequency of countwheel chiming in the eighteenth and nineteenth centuries, once rack chiming was an established alternative. It is likely that the full rack system on both striking and chiming sides was the more expensive arrangement; certainly it is the one discussed by writers of these periods. On the other hand, countwheel chiming certainly continued quite unaffected by the virtual demise of ordinary British countwheel striking in quality clocks. Even by modern standards, and certainly by those of fifty years ago, quarter chiming on the run was an abnormal luxury. It was preferred for functional rather than decorative use and therefore it was extensively developed in the large variety of pull-repeat systems in bracket clocks which did not chime in passing; in fact, it must have been intended mainly for night use. The commonest chimes were ting-tangs and four- or six-bell scales which, kept in sequence by the rack system, were suitable for repetition work.

Plate 51 Eighteenth-century calendar movement with centre seconds and external countwheel chiming on two bells. The wheels above and to right of centre pipe are for a year calendar with central hand. One wheel (identical to that at the top) meshes with the worm-driven wheel and is on the central axis for the calendar hand. This has been omitted to show the displaced train which is necessitated by centre seconds

Plate 52 The chiming and striking levers visible once hour and calendar wheels are removed from the calendar movement (Plate 51). Note hour let-off lever from pin in countwheel. The countwheel detent is connected to pin-locking mechanism of the chime (see also Fig 23)

Plates 51–4 depict a large, somewhat elaborate movement which may date from about 1760. It has three-train ting-tang chiming, centre seconds arranged so that the chime-winding square has to be placed asymmetrically, an annual calendar with a chapter-ring having 365 divisions outside the minutes, and a moon-phase indicator. The calendar mechanism – the large wheel at the top with a pin for the moon phase, the lantern and worm gears, the smaller wheel for the worm, and another large wheel which has been omitted because it obscures the countwheel – largely conceals the sounding arrangements and so its wheels, too, have been omitted. There is a rack system and starwheel snail visible on the left for the hours, and a countwheel with ratchet-shaped notches on the right for the chimes. Incidentally, according to Grimthorpe, a musical interval of a fourth is best kept for ting-tang and these bells in fact conform. However, a third, as in cuckoo clocks, seems just as common.

The lifting piece can be seen to be worked by the equivalent of the cannon pinion with four equally spaced pins in the centre, there being no central minute arbor in the train; and the starwheel snail to be

94

Plate 53   Details of hammers, small pinwheel and chime warning piece of the calendar movement (Plate 51)
Plate 54   Chime locking, ting-tang pinwheel and hammers of the calendar movement (Plate 51)

driven by the equivalent of the minute wheel just below. This lifting piece is angled and pivoted in the middle, and the horizontal top section goes through the front plate to meet the chime warning wheel. The chiming is locked by the pin method, and the long detent descending to the countwheel is attached to the locking-piece arbor so that locking and a countwheel slot must coincide, in the usual way (Fig 23). The chime hammers are activated by a pinwheel of six pins driven by the chime intermediate wheel.

At the hour the central lever, which is equivalent to a modern strike flirt and is provided with a knop for possible repetition of the hour, comes into play. The countwheel has a pin which is so placed that it engages the tip of this lever when the chiming is in the fourth quarter, and the tip is shaped to set off the strike warning and lift the strike rackhook, with the usual clearance. It will be seen, from the clockwise rotation of the countwheel, that the shaping of this tip is very critical and, in particular, that it must be just long enough to release the strike warning as the chiming comes to an end. Contrary to later practice, the strike warning piece comes into position only on the hour.

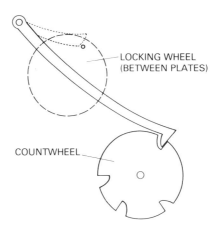

LOCKING WHEEL
(BETWEEN PLATES)

COUNTWHEEL

Fig 23   Countwheel chime locking (eighteenth-century)

This is a simple and very reliable system, although in the nineteenth century such a chime would certainly have been arranged by the previously rare quarter striking, the only audible difference being that the system illustrated strikes four ting-tangs at the hour whereas quarter striking could not. But reliable as it is when once set up, the movement presents problems in assembly. The chime hammer assembly is a detachable unit and it is difficult to put the movement together with it in place; on the other hand, these hammers can stop the chime if they are on their pins when the train is at rest. Therefore one can only guess where they will be as one assembles with the train locked and then, on being wrong, separate the plates and rearrange the locking and warning which in fact allows very little run if the bells are to be clear. There is similar difficulty with the countwheel. This must have the shaped detent properly in one of its shaped slots when the train is locked but, as it is pinned to a squared shaft outside the plate, this is not easily arranged when the wheels are being assembled. It was not until the twentieth century that these external countwheels were customarily fitted with set-screws so that they could be positioned anywhere on their arbors – a vast improvement.

Countwheel chime came into its own in the later nineteenth century, by which time somewhat pretentious cases were being fitted with relatively cheap mass-produced movements enriched by the melodious booming of tape-steel gongs, perfected since their first appearances in clocks, probably in the 1820s. Very soon these were generally superseded by the softer and lighter rods of phosphor-bronze which were introduced in about 1880. For most of the century the bracket clock with chiming was still a luxury item and the rack system was generally followed. These marvellous, but in a way outdated, productions of firms like Thwaites and Reed of Clerkenwell were overtaken by

96

a flood of French, German and American clocks which, because of their price, reached a far larger public. A leader among European Continental exporters was the firm of Arthur Junghans, whose name was behind many a British-looking walnut director's or presentation clock at the turn of the century. Junghans had pioneered mass-production in Germany, having learned much from the USA, and for the next fifty years or so the firms of Junghans and successors Lenzkirch, Becker, HAC (Hamburg American Clock Company) and Winterhalder and Hofmeyer had a large share of the British market. Their more elaborate clocks display a very wide variety of striking and chiming mechanisms based on the countwheel.

Plates 55–8 show a movement by Gustav Becker of about 1890 and we will compare this with a movement by Junghans known to have been a presentation piece in 1907. The former, at first sight, seems rather strange and complex. Becker, who started out as a one-man businessman, ended by taking over clock manufacture in Freiburg and his firm which merged with that of Junghans in 1926 managed to combine considerable innovation with mass-production, and in fact played an important part in the development of the then new '400 Day' clock. The oddness stems from the fact that the long lifting piece, which stretches right across the clock from the manual let-off on the left to the usual warning slot on the right, not only lets off the chime but also controls the hour rack. Further, there are two hour pallets, one for gathering and one for locking.

At the quarter, the lifting piece, which has a one-way detent permitting the hands to be reversed, raises the right-hand forked lever, which also has manual let-off, and its bent tip inside the movement sets the warning (Fig 24). The forked lever has a pin mid-way (visible as a dot), which proceeds through another concealed slot into the movement, where it becomes the countwheel detent. On its arbor is a locking piece, adjustable by a set-screw fixing, which engages with a pinned locking wheel. Thus the raising of the lifting piece sets the warning and, if properly set up, releases the countwheel and locking wheel so that the train can run to warning, which is released in the usual way when the lifting piece falls. The forked locking and detent piece has at the end of its top branch another long pin into the movement, and this engages with a freely pivoted lever, as is typical of chiming clocks of this date, to set the strike warning whenever chiming takes place. But although the strike warning piece moves into position, the strike train is not released during the quarters.

At the hour, the fourth lifting pin is, as becomes normal in later clocks, set far out so as to raise the lifting piece extra-high. The effect of this is to cause the pin on the top branch of the lifting piece to move up the chime rackhook (the Y-shaped piece pivoted at the top near the

Plate 55   Chiming movement by Gustav Becker, c1890, showing hour warning lever operated by an extension of the chime countwheel detent, hour rack released by extensión of lifting piece and higher lifting pin

Plate 56   Internal countwheel, locking piece and hour warning piece of the Becker movement (Plate 55)

centre) and release the hour rack with its extension to the left which, by a pin and the curled pallet there, locks the hour train. The lifting piece and rackhook are so shaped, and the pins so adjusted, that at the hour the chime warning is set, and then the hour warning, before the chimes start. The hour rack is released shortly afterwards, but of course the train is held by the warning piece and this is not released until the chime-locking lever on the right, which operates the pivoted warning piece, falls back because its pin can enter the countwheel slot after the quarters. The hour-rack gathering pallet runs on a pinion between plates and has twice the number of teeth as has the locking pallet. This is because the gathering pallet is double-toothed consisting of two pins projecting from one steel head, and it gathers two rack teeth with one revolution. It is set up so that the pins are as far out of the rack as possible but, even so, they obstruct the rack's fall. Therefore the pallet runs in a large hole and rests by gravity in the rack teeth; when the lifting-piece arm releases the rackhook, it simultaneously pushes the pallet out of engagement with the rack so that the latter is free to fall.

98

Plate 57   Two types of centrifugal fly on the Becker movement (Plate 56) – one with moving weight, the other with moving blades. Chime 'barrel' is composed of pinwheels screwed together and contains all ten sequences
Plate 58   Detail of chime pinwheels and early form of link to hour hammer on the Becker movement (Plate 55). Silencing is by pressure on hammer tails

Much thought has gone into the design. The levers are carefully counterbalanced, as is the rack. The locking is adjustable, the lifting piece is 'safe'. Two sorts of fly have been selected for the different trains, and there is friction adjustment to the beat. Yet the result seems needlessly complicated, with considerable difficulty for the repairer should levers or pins be bent. The setting up of the strike locking, warning and gathering is tricky; the chime locking piece is adjustable for height, but the countwheel and locking wheel have still to be related to each other during setting up, for neither is movable on its arbor. The barrel substitute – four discs with fine projecting pins screwed together – was common at the time and no doubt economical; but should those discs be separated, for example to replace a missing pin, they are virtually impossible to reassemble unless marked carefully and set out in order at the time. Perhaps there is some consolation, however, in that they are simple 'Westminster' and moreover – since they are mounted directly on the countwheel arbor without ratio wheels, and thus revolve once instead of the usual twice an hour – they have a set-screw for adjusting them to the countwheel

99

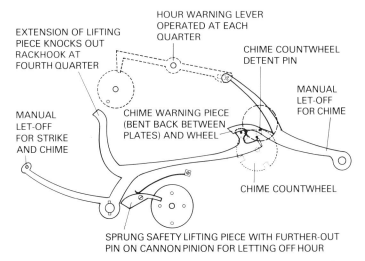

HOUR WARNING LEVER
OPERATED AT EACH
QUARTER

EXTENSION OF LIFTING
PIECE KNOCKS OUT
RACKHOOK AT
FOURTH QUARTER

CHIME COUNTWHEEL
DETENT PIN

MANUAL
LET-OFF
FOR CHIME

MANUAL
LET-OFF
FOR STRIKE
AND CHIME

CHIME WARNING PIECE
(BENT BACK BETWEEN
PLATES) AND WHEEL

CHIME COUNTWHEEL

SPRUNG SAFETY LIFTING PIECE WITH FURTHER-OUT
PIN ON CANNON PINION FOR LETTING OFF HOUR

Fig 24   Chime lifting and warning on movement by Gustav Becker

sequences. The chime is not, at this date, self-correcting, and this is the reason for the manual let-off facilities. Since the hour can be struck with the hand only in one position as decided by the eccentric lifting pin, the quarters have to be let off until four quarters have been struck when it is in that position. These will automatically be the right quarters if the barrel has been put on correctly with its first quarter corresponding to the first cut-out on the countwheel.

Two other minor characteristics of the time are worth noting. The first is that the strike/silent lever, seen above the movement since its indicator is in the break-arch, is on a friction-tight arbor and lowers a bracket onto all the extended hammer tails to raise them out of contact with the chime-barrel pins. The second is that two of the winding squares are geared – a device adopted not primarily to make winding easier but to arrange for the actual squares to be symmetrical inside the chapter-ring. This is a common source of trouble, for with the large forces involved in chiming trains these gears wear and eventually lose some teeth. There is then no option but to cut a new wheel, or to have one cut.

The chiming of this movement is on rods, not very long after their introduction. They are screwed into a stout cylindrical, but not hollow, block, and this is mounted on a separate soundboard screwed to the case. In the Junghans movement of 1907 (Plates 59–61) the same diagonal position from top to bottom is used for the chimes, but they are screwed into a hollow cylinder which is firmly bolted direct to the case – a more vibrant arrangement. Both movements have similar rotating clamps for the chime rods, and indeed such things continue to the

100

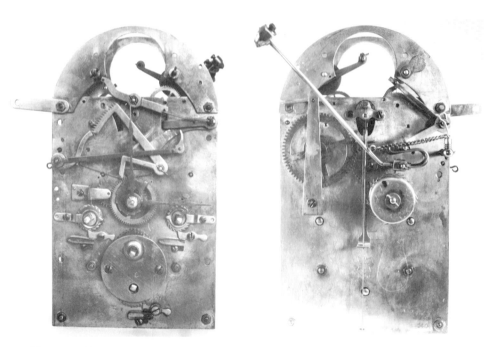

Plate 59   Chiming movement by Arthur Junghans, 1907. Note chime pallet operating lever to release hour rack, also the prominent pivoted hour warning piece, raised by a wire lever on the chime warning piece arbor

Plate 60   Rear view of the Junghans movement (Plate 59) showing early form of hour-hammer link (cf Plate 58). Note solid 'snagged' barrel and silencing by pressure on hammer tails. The arrangement of hammers on the hour side and resulting ratio wheels are unusual

Plate 61   Detail of internal countwheel, locking piece and locking cam on the Junghans movement (Plate 59). Countwheel detent and locking piece are on the same arbor

present day for chime rods are, at their junction to their screws, very fragile. In the Becker movement, and more typically in later arrangements, the hour hammer is linked by a lever across the back plate to the hammer-tail arbor. In the Junghans movement the hammers are arranged by ratio wheels on the strike side, which is unusual, and it will be seen that the hour hammer is linked by a little chain to the hook coming from the tail arbor. The hammers of both movements are brass with leather inserts.

We are still in the era of the somewhat inconvenient internal countwheel which so complicates setting up, of geared winding and of strike/silent in the break-arch – partly a reproduction feature. Clocks even before the quarter striker in Plate 42 of about 1900 had what was to become, at least for the cheaper range, typical strike/silent 'switches' brought out through the dial at three o'clock and nine o'clock (change of chime). The mere unlabelled tip of a strike/silent lever at twelve o'clock, without indicating hand, was quite common in the eighteenth century. But the Junghans mechanism is a good deal simpler both to follow and to set up than the Becker movement. It employs a typically European Continental brass hour rack and rackhook; the rack tail has no safety device. The snail screwed to the hour wheel is remarkable only in that it has the cut-out which ting-tangs and quarter strikers require in the one o'clock section. There is no need for it here, so presumably it is due to some feature of factory mass-production; small wonder that the owners know the clock as the one that strikes 'three at one'!

Despite typical features, there is still clearly nothing like a settled form of countwheel chime mechanism such as later developed. The safety-angled lifting piece lifts a long flirt-shaped lever attached to the warning piece but loose on its arbor. The connection with countwheel and locking is by the wire lever fixed to the end of this arbor, it being able to be raised by a pin in the warning piece (Fig 25). On this same arbor, between the plates, is a combined locking piece and countwheel detent, working on different wheels (Plate 61). This is not unlike the Becker locking piece, adjustable on its arbor for height, and it works on a hoopwheel-like cam, catching from behind the vertical face.

Here, as in the Becker movement, the locking piece and detent, by means of the wire lever attached to their arbor, tilt a centrally pivoted lever whose other end sets the hour warning at each quarter, though the train moves only at the hour. The release of the rack is, however, totally different and very much simpler, being achieved by a hooked pallet on the countwheel arbor which raises a long lever attached to the rackhook. The rackhook is attached to a locking piece which works internally on a locking-wheel pin in typical Continental fashion; but, again, this locking piece catches, rather than receives, the pin. The gathering pallet is simplicity itself, being merely a pin in a small brass

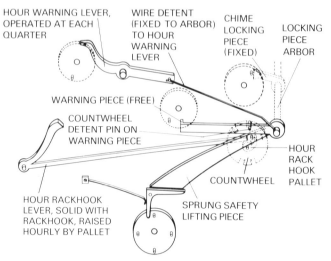

HOUR WARNING LEVER, OPERATED AT EACH QUARTER

WIRE DETENT (FIXED TO ARBOR) TO HOUR WARNING LEVER

CHIME LOCKING PIECE (FIXED)

LOCKING PIECE ARBOR

WARNING PIECE (FREE)

COUNTWHEEL DETENT PIN ON WARNING PIECE

HOUR RACK HOOK PALLET

COUNTWHEEL

HOUR RACKHOOK LEVER, SOLID WITH RACKHOOK, RAISED HOURLY BY PALLET

SPRUNG SAFETY LIFTING PIECE

Fig 25   Hour release, chime lifting and warning on movement by Junghans

blob. The hour hammer tail works on a starwheel; that in the Becker clock worked on the older-fashioned pinwheel. The strike/silent lever arranges for the hammer tails – in fact of the chime only – to be pressed down as in the older movement, and geared winding will be noticed on the going, not the strongest, spring.

This is a very much simpler and more robust movement than Becker's and both clocks' sounding repertoire is the same. However, there are several indications of the shape of things to come, besides those already noted, in the Junghans movement. Perhaps most importantly, it employs external ratio wheels on the back plate. These are not simply so that the hammers can be taken over to the right, perhaps to shorten the hour-hammer linkage, but to impose a 2:1 ratio between countwheel and barrel – for this chime barrel, as was to become universal practice, revolves twice in an hour. Significantly, also, it is composed not of fragile pinned discs, but is a steel cylinder with snagged teeth as pins. Admittedly, this type of barrel is difficult to repair and soldering cannot be avoided, but it is a great deal more robust in the first place and there is no possibility that the 'pins' may go out of order when the clock is dismantled. One other detail may be mentioned. The movement in Plate 12 has as large a fly as space permits between plates, and resorts to a cut-out to clear the pillar. We have seen two different types of sprung centrifugal fly in the Becker clock, and now, in the Junghans, we have the real solution when a large fly is required to slow a chiming train, especially with gongs; namely to take the arbor through the back plate and fit a large fly externally with its own supporting cock, as indeed had long been the

Plate 62 German external-countwheel chiming movement of the 1920s with self-correction by clip on countwheel (see page 106). Chiming warning and locking slots clearly visible. Strike flirt incorporates countwheel-detent. Pin and cam pallet locking on hour rackhook. Silencing by raising lifting piece

practice in superior, and especially musical, clocks. Plate 45 shows another example, and we shall see even larger ones in lavish rack-chiming movements.

None of these countwheel-chiming movements has been self-correcting. That featured in Plates 62–4, of German origin, is typical of the earlier mass-produced self-correcting chiming clocks which became popular between the boom of the 1920s, the crash in the early 1930s and World War II – a time when it became increasingly desirable and fashionable for the middle classes to own their own house and to have a chiming clock on the mantel of the parlour or living room. Perhaps the most far-reaching development was to place the controlling countwheel on the outside of the front plate, where it was adjustable on its arbor by set-screw. This enormously simplifies assembly, checking and correction. The cases generally were of plywood, often in the 'Napoleon's hat' shape, but later more box-like; the chimes were initially parallel with the base of the clock with the hammers on the back plate, but were later often underslung, the whole assembly being below the plates to lessen the depth of the clock. Chimes almost always included Westminster, whilst the more expensive movements had a pump-change to one of the several variations of Whittington chimes or St Michael, on eight gongs. The width of the eight gongs virtually compelled adoption of the underslung arrangement. Similar movements, sometimes with balance wheel and sometimes with short pendulums, were fitted into plywood 'grandmother' cases, where longer rod gongs and deeper notes could be used. Whilst weight-driven quality longcase clocks continued to be made with traditional movements, these cheaper clocks were generally spring-driven and probably the majority have, as in this example, separately detachable mainspring barrels. This the modern repairer appreciates; the forty- to fifty-year mainspring life of these clocks, particularly on the chiming side, is now exhausted, and a good proportion of repairs concerns broken or 'dead' mainsprings for which replacements are readily available from the suppliers.

Chiming is let off by a starwheel cam fixed to the cannon pinion. One tooth of the four is usually longer than the others and is used, as we have already seen with an eccentric lifting pin, to distinguish the hour. The long curved lifting piece has on its arbor a locking piece, fixed by set-screw, which meets the locking-wheel pin although in these clocks this is not responsible for locking, but for warning. The train is locked by what would usually be the warning piece, projecting back through the slot and resting on a small hoopwheel-type cam. This locking piece is bent back from the long lever which rests on the lifting piece which has, by a similar projection forwards, the function of countwheel detent; thus locking and countwheel control are, as so often, by the

Plate 63   Note how conventional form of link to hour hammer, single ratio wheel and cam form of chime barrel on the German 1920s movement (Plate 62)
Plate 64   Side view of the German 1920s movement (Plate 62) showing adjustable warning piece acting on 'locking' wheel; to this wheel's arbor is attached a cam on which the locking takes place. There is no warning wheel as such

same double lever, but locking is on what would normally be the warning wheel. This lever is variously named, but we will call it the 'strike flirt'. The adjustable warning piece on the lifting lever is set up with a very short run, so that it is in the wheel's path when the strike flirt releases the cam locking, and the train runs when the lifting piece drops off the lifting tooth at the cannon pinion.

The strike flirt is so called because, when it is raised at each quarter, its tip, which is the strike warning piece, moves to warning position, though the strike train is not released. The high fourth-quarter section of the countwheel raises the lifting piece and strike flirt extra-high so that the strike flirt not only warns, but also raises the strike rackhook, letting the train run to warning, then run free, with rack released, when the strike flirt falls at the end of chiming – the fall being caused by encountering a countwheel notch. The rack striking is of the typical modern type already considered, with pin gathering and pallet-cam locking. The 'silent' lever to the right shows how far mass-production and international trade have gone, for it is so arranged as to be moved in and out to suit different cases. It silences chime, and therefore also strike, by raising the lifting lever away from the lifting star. The drawback to this arrangement is that it sets the chime ready to perform as soon as it is switched on again, when it will most probably be at the wrong sequence for the time shown by the hands – as of course may also happen if the hands are carelessly adjusted. To cope with these eventualities, self-correction is arranged.

The principle to remember when dealing with the wide variety of self-correcting devices found, is that they depend on the fact that one quarter – the fourth – is different from all the others in that its lifting tooth is extra-long and so the chime sequence is held at the third quarter until the long lifting tooth comes round. A secondary locking device is added to hold the chime silent at the end of the third quarter, and only the extra lift given by this tooth will release it. In the present case a simple and very popular arrangement is used. It consists of a small sprung hook attached to the back of the countwheel in such a way as to engage a pin in the lifting piece where it passes behind this wheel. The locating pin of the hook is visible on the front of the countwheel in the small slot in which it moves as the sprung catch comes to a stop against the lifting-piece pin. If the hands do not indicate that the next chime required is the hour, the lifting piece will not be lifted high enough to free the catch and pin; only when the big hour lifting-tooth comes round is the lifting-piece pin lifted out of the catch and the countwheel allowed to revolve for the hour. This is a simple, well-proven and reliable device, and it poses no problems in assembly since it does, indeed, sort itself out.

The hammer and barrel arrangements adopted here are such as

became very general. The hammers are in a sub-unit – often, but not here, removable as a whole – on the back plate, and the barrel is driven by one ratio wheel with twice the number of teeth as are on the barrel pinion. This ratio wheel is mounted with a set-screw so that the barrel can be turned to set the sequence to correspond to that registered by the countwheel. The barrel itself is composed of four steel toothed discs with spacers screwed together. These discs have to be watched, but they are more robust and do not present the same potential confusion as the full-chime discs with pins seen on the Becker movement. For silencing of chime, but not hour, the usual friction lever is present; but it no longer acts on extended hammer tails, simply raising the hammers themselves. The hour strike in this instance employs one separate hammer and two of the chime hammers, and is contrived by a pivoted linkage whose tail enters the back plate and engages with the hour starwheel. The starwheel almost universally replaced pinwheels at this time in mass-produced clocks; it was no doubt cheaper to make and is, besides, very much stronger. The damping spring is somewhat unusual; reliance is usually placed on the hammer-tail springs and the resilience of the long hammer-shanks themselves.

Such mass-produced movements with stamped parts of thin metal, but in higher grades polished steel levers and frosted or grained back plates, are found in cases varying from the pretentious to the frankly utilitarian into which the movements can be very difficult to fit, so limited is the space. The cheaper ones fasten the relatively light movements by means of lugs screwed onto extended pillars and fitted with too-short screws into the thin plywood. Better and earlier ones, as here, employ some kind of seatboard with screws up into the bottom pillars of the movement. Strike/silent and manual release of the countwheel are usually provided. The manual release is not of great assistance with self-correcting movements and, though it lets off the strike as well as the chime regardless of quarter, it is of no use as a repeat device since it advances the chiming by a quarter. In the better models (Plates 73–8), changes of chime were produced in the hallowed manner by means of a lever which pumped the barrel in line with the hammers for a different chime. But this was usually done by a cranked lever at the side of the dial and passing round to the back of the barrel, rather than by the straight lever passing over an arbor with an angled end; and my impression is that it was largely confined to underslung barrels and hammers – no doubt because of the greater depth required. It was general to employ some form of adjustable crutch. Regulation from the pendulum's suspension, as here, was less common, except in quality reproduction movements.

There are no special problems in setting up such movements. The adjustable warning piece has simply to be set so that it is in place before

107

Plate 65   Classic Enfield external-countwheel chiming movement of the 1930s. Note typical crossed-out countwheel and parallel lifting-lever and strike flirt. For self-correction – one of several Enfield systems (see page 111)

Plate 66   Rear view of the Enfield movement (Plate 65)

Plate 67   Detail of self-correction system on the Enfield movement (Plate 65)

the chime train is released, and the countwheel can (at last) be suitably positioned with the train assembled. Because the warning is so early in the train, the usual half-revolution's run would be completely excessive and could well cause partial chiming on warning; thus, when the train is locked at the wheel before the fly, the warning pin should have only some 5mm (³⁄₁₆in) to run. This must, of course, be arranged as the wheels are being assembled, although the detachable barrels do facilitate subsequent adjustments. The hour train warns, as is more usual, on the wheel before the fly – the warning wheel; and half a revolution is reasonable here.

Enfield movements, produced in very large numbers, were of higher quality and finish; the plates were frosted, the steelwork blued or polished, solid adjustable pallets were used, oil-sinks were put in, and a number of adjustment points incorporated. Plates 65–7 show a typical Enfield four-gong movement, from a plywood 'fumed oak' grandmother case, and after fifty years it is very little the worse for wear. (Identical movements were produced until very recently, but with 'floating balance' escapements.) It will be seen that the standard layout is followed and the striking arrangements are very little different from the previous example, but the chiming and self-correction mechanism

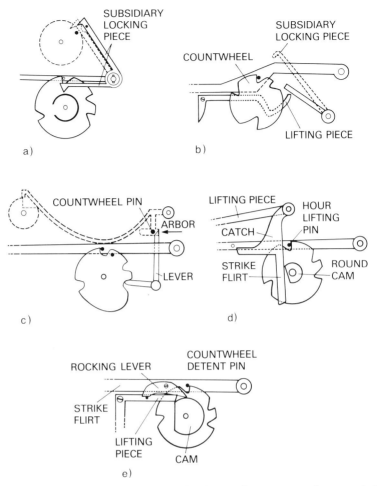

Fig 26 Other self-correcting devices (see also Fig 25) shown at second quarter before coming into effect

(a) Subsidiary locking piece

(b) Early form; unusual-shaped lifting piece frees subsidiary locking imposed by deep slot in countwheel

(c) The arbor is sprung so that it can be moved to the left by the lever, which is done by the countwheel pin. Only when this happens (when the countwheel detent is in the end of third quarter notch) does subsidiary locking take effect, and only extra lift at the fourth quarter can free this locking

(d) The catch is freely attached to the lifting piece and lifts the strike flirt and detent so long as it rests on the round cam; when it meets the flat side, the catch cannot lift the strike flirt until there is extra lift from the hour lifting pin

(e) Rocking lever principle. There are several forms. The lifting piece can only raise the strike flirt and countwheel detent when the lever's tip rests on the rounded cam. When the flat face is presented, only a high hour lift can free the strike flirt

110

are far from identical. There is little protection in the previous movement if the hands are pushed backwards even though the long lifting piece is bent to be diverted by the cam, whereas the Enfield employs what became a very general device – a short, sprung, right-angled lifting piece which can be moved either way. Directly bearing on this is a short pivoted warning piece, which works through the plate in traditional fashion on the pinned wheel next to the fly. Locking is by an adjustable hook caught from behind by a pin on the next, locking, wheel – the right-hand of the two such hooks visible in the side view. The hook is mounted by set-screw on the same arbor as the strike flirt which, as before, contains the countwheel detent (a pin) and the strike warning piece, and which is raised, after the appropriate clearance, by a pin projecting from the chime warning piece. Thus, as a quarter comes round, the warning piece is lifted into position and then raises the strike flirt, unlocking the chime train which runs to warning and runs freely, subject to the countwheel, when the warning piece falls. At the hour, all this occurs, but directly beneath the countwheel-detent pin on the strike flirt there is a projection designed to be raised by a pin in the fourth quarter's section of countwheel, and this causes the strike flirt to move so high that it knocks out the strike rackhook and leaves the strike at warning until the countwheel allows the flirt to fall, when striking begins. Thus the Enfield's traditional countwheel pin replaces the raised countwheel section – probably the commoner modern arrangement – for letting off the hour strike.

Again supplementary locking is employed at the end of the third quarter to produce self-correction, and again it can be released only by a longer lifting tooth for the hour. Next to the locking hook on the strike-flirt arbor, an identical second hook is loosely mounted, held raised by a pin on the first (Plate 67). This second hook has, attached to it, a lever which is bent to pass out from the front plate and rest on a cam which is fixed to the back of the countwheel and which has a deep notch at the end of the third quarter's countwheel section. The lever falls into it at that time and so the second hook falls alongside the locking hook on the locking pin and double-locks the chime train. With the usual extra lift at the hour, the locking hook moves higher and takes the correcting hook with it, so freeing the train; at any other quarter when correction is in force, the strike flirt does not rise high enough for the locking hook to raise the supplementary hook. The system has no snags in assembly, so long as the hooks are kept to the right side of the locking pin, and is entirely efficient.

Great ingenuity was devoted to self-correcting devices at the time and this is not the only one used even by this maker. Some other systems are shown in Fig 26, the ultimate aims doubtless being simplicity, reliability and economy. In my view, the arrangement with

Plate 68 Contemporary underslung chiming movement with external countwheel; showing minimal countwheel and a free duplicate locking piece as self-corrector (see page 110)

a catch on the countwheel is robust and reliable and the alternative systems do not have any clear advantages; it is evident, however, from clocks one has seen, that they prove problematic to some repairers. There really is no reason why this should be so, provided that the principle of supplementary locking after the third quarter is kept in mind.

Finally, the contemporary Westminster movement in Plates 68–9 shows both what changes of detail have taken place, and how little the outline, developed between the wars, has changed. Typically of an underslung chimer, there are three ratio wheels, which are merely duplicates in the gearing to get the drive down to the barrel. The chime barrel, though not visible, consists of cheaply produced ganged discs, now no longer separable. The hour hammer linkage raises a bass gong hammer and two of the chime hammers. The movement is secured by lugs, as was the Enfield movement, in a case now of synthetic – rather even than ply – wood. The clock is given a floating balance; in the 1930s, many were provided with lever platform escapements, and this is particularly common in grandmother clocks which therefore gain

Plate 69 Rear view of movement in Plate 68, showing use of floating balance escapement (less affected by level) and range of ratio wheels needed to reach underslung chime barrel

nothing but appearance, and longer gongs, from their height. The basic layout of countwheel chime, strike flirt and rack strike remains; but the countwheel is much reduced in size, a sprung hour rackhook pivoted at the bottom is employed and the rack, as is now general practice, is pressed from solid sheet metal. Locking and self-correction, but not warning, all take place on the front plate. The barrels are for some reason (probably economic) not detachable, and they will prove a nuisance in years to come.

The cranked lifting piece passes beneath the hour rack and is operated by a snail rather than a starwheel, but still with one section substantially higher than the others. Its end – to the right of the disc next to the rack – is bent round into the movement as warning piece, and another branch has a projecting pin which raises the strike flirt. The disc referred to is in fact a locking pallet with a pin, which catches on a vertical face of the strike flirt to lock the train. There is the usual clearance between warning and the pin which raises the strike flirt to free the chime train. The strike flirt bears a projecting pin as countwheel detent to the tiny countwheel which is adjustable, being set

113

by a grubscrew not visible in the illustration. The strike flirt continues as usual to the left, to become the strike warning piece; and the countwheel has a clear bump at the fourth section which raises the strike flirt so far that the pin close to the warning piece knocks out the rackhook by means of an extension of the latter, not visible here, behind the rack. The striking reflects the chime movement in Plates 65–7, in that it is locked by the extended rackhook on the wheel before the fly – normally the warning wheel – and warning is given by what would normally be the locking wheel. It follows that this wheel requires much less run to warning than is usual.

The self-correcting device is, yet again, different from others noted so far, though similar in principle. The countwheel has, fixed behind it, a cam with a notch at the end of the third quarter on which rides a detent pivoted freely to the same stud that carries the strike flirt. This detent is simply a duplicate of the first part of the strike flirt, ie a hook which catches the locking pin at the end of the third quarter. When the true time for the hour arrives, the high lifting cam raises the strike detent, via the locking piece and pin, so high that the countwheel-detent pin raises also this second locking piece, and the train can run. At any other quarter, the lower lifting action determines that the additional detent cannot be raised, and the chime train waits.

This clock is, then, in an electronic age, something of a relic of the largest expansion of chiming clocks that has ever occurred, for it is not a reproduction but a cheap chimer for the modern living room, and it is the final phase of the countwheel. The system that began by controlling the hours and quarters in non-domestic clocks of the Middle Ages, continued in use for hours and quarters until the turn of this century and was finally almost abandoned for hour striking, culminated as the standard controller for quarter chiming in the last age of the mechanical clock. Now it is employed, as increasingly are mechanical clock-workings, more largely in reproductions in traditional wooden cases. However, before it fell into disuse, it evolved, with self-correction, to a more or less standard form which had reached perfection. If it had not been superseded by electronic devices, one could envisage no further development for it.

6

RACK CHIMING

Although the layout varies and there are two let-off systems, rack chiming has altered very little since it became probably the most used system of quarter chiming in the later eighteenth century. Because it cannot go out of sequence it has clear advantage over the countwheel system. On the other hand, it is very much heavier, more subject to adjustment and to wear, and more expensive to produce. Nineteenth-century writers cite it as a matter of course as the arrangement for quarter chiming; but, as we have seen, it was being superseded, partly for economic reasons, towards the end of the century and, from perhaps 1880 onwards, it was progressively reserved for the most expensive and lavish clocks, and was appropriately styled and finished.

Fig 27 outlines the standard form of rack chiming and Plate 70 shows a reproduction movement using the same parts in a somewhat different layout; such variations are very common, particularly in the eighteenth century. Both are of the non-repeating set-up which employs a warning in the chiming train; in this arrangement there may or may not be a starwheel-mounting for the snail – certainly it is not needed.

It will be seen that lifting is from the minute wheel with four identical pins in the rim, and that to this wheel is fixed a quarter snail. The wheel revolves anti-clockwise and raises a cranked lifting piece to whose tip is attached a warning piece going into the usual slot in the front plate. The lifting piece also raises the rackhook to the vertical chime rack on the right; it may do so by a pin on the warning arm of the lifting piece or by a detent pivoted on the same stud. In the eighteenth century the lifting pins were often on the cannon pinion and so the lifting piece was pivoted to the left and cranked to reach over to the warning piece and quarter rackhook beneath which it came and which it raised directly.

The working is identical to that of rack striking. The lifting piece rises until the train is held at warning, the rack falls, and the pallet gathers it up and locks the train; pallet-tail locking is almost universal, but locking by a piece on the rackhook arbor and a pinned wheel also occurs. Each tooth of the rack corresponds, in the design of the train, to one sequence of chiming and the rack normally has 5 or 6 teeth; in fact

115

STRIKING   CHIMING

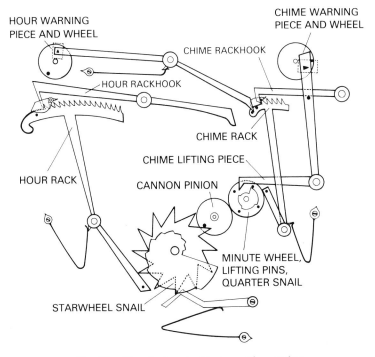

HOUR WARNING
PIECE AND WHEEL

CHIME WARNING
PIECE AND WHEEL

CHIME RACKHOOK

HOUR RACKHOOK

CHIME RACK

CHIME LIFTING PIECE

HOUR RACK   CANNON PINION

MINUTE WHEEL,
LIFTING PINS,
QUARTER SNAIL

STARWHEEL SNAIL

Fig 27a   Standard rack chiming with warning

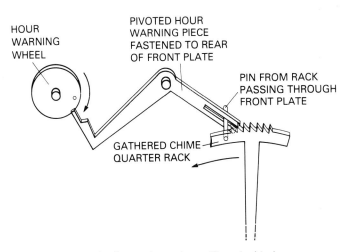

HOUR
WARNING
WHEEL

PIVOTED HOUR
WARNING PIECE
FASTENED TO REAR
OF FRONT PLATE

PIN FROM RACK
PASSING THROUGH
FRONT PLATE

GATHERED CHIME
QUARTER RACK

Fig 27b   Internal warning with rack chiming

116

it must have at least 5, since there is a tooth between pallet and rackhook which cannot be gathered by the pallet. The chime snail, like the chime countwheel, is based on cumulative sections within a total of 10 divisions, as 10 sequences are required in an hour's chiming.

The chime mechanism lets off the hour striking, and this is usually the most troublesome area of these movements. The hour warning piece is the end of a long pivoted lever, the hour end sprung to stay up, the chime end held up, against the spring, by a pin in the end of the rack, either in front or behind. Thus whenever the chime rack falls, the hour warning piece rises into warning position, but, as the hour train is locked, it simply returns down when the rack is gathered again. However, the hour rackhook is controlled by the fall of the chime rack which, only at the fourth quarter, falls far enough to knock it out, releasing the hour train which then is held at warning until the chime rack is gathered up. Once the chime rack is gathered, the hour warning is released and hour striking begins.

Both the hour warning and the hour rackhook and spring call for nice judgement and need attention from time to time. The hour warning piece, like an hour rack tail, is usually joined friction-tight at its stud, so that the detent engaging with the rack can be adjusted relative to the position of the warning piece in the slot to the left. The warning piece must not jam on the bottom of the slot so that the final gathering of the rack is uncertain, but equally the warning piece must be fully in a position to arrest the train by the very short time that the quarter rack has knocked out the hour locking. If the warning does not hold long enough, the strike may begin with, or too close on, the chiming. It is therefore necessary for the strike warning blade, which the warning pin acts upon, to engage the pin throughout the gathering of the last tooth of the quarter rack; the blade is therefore quite large and arranged to catch the warning pin at the bottom, so that the pin can slide the whole way down it as the chime-rack tooth is gathered.

If the quarter-rack spring is too stiff, plainly there can be difficulty in gathering; on the other hand, if it is not stiff enough, the rack will not fall sharply enough to knock out the hour rackhook. The latter also is usually adjustable at its pivoting point so that the fall of the quarter rack ensures freeing of the hour train but, at the same time, there is no risk that this will occur with the shorter fall at the third quarter. These matters have to be attended to during setting up along with such things as ensuring that the train is released when the hand points to quarters and that the starwheel snail, if any, moves only just before the hour; but they are all external, on the front plate. Although the warning run must be set at a quarter of a revolution, and the chime barrel may have to be adjusted so that the hammer tails are off their pins when at rest (for on older clocks the barrels are often internal) the

Plate 70 Rack chime with warning in a reproduction movement by Classic Clocks Limited. Right-hand forked lever, by a linkage, moves internal barrel in and out for Westminster (eight bells) or Whittington (four bells) chimes. Three fusees control the springs

problems of setting up of rack chiming are less than with countwheel – particularly internal unadjustable countwheel – chiming.

Such is the system with warning, which appears to have been commonest during the eighteenth century. Towards the end of that century the alternative flirt-release system became increasingly popular. Rees, early in the next century, illustrates rack chiming with flirt release as if it were the normal thing, with a starwheel snail and repeating lever. Britten, in early editions of his *Watch and Clockmakers' Handbook*, describes the arrangement with warning and interestingly – apparently not thinking of repeating – observes that 'this is clearly a much better arrangement than the usual flirt, which absorbs more power and is less certain in its action'.

The movement by John Ellicott illustrated in Plates 71–2 is of distinguished provenance. It is something of an oddity (probably having been altered) but it shows a typical early form of flirt release. It

118

Plate 71   Rack-chiming movement by John Ellicott, *c*1780, with flirt release. Strike/silent lever (by nest of bells) limits flirt's movement. Double-toothed rack of Continental style; distinctive hour warning system by spring-loaded rack tip; distinctive hook for pin locking of hour strike on 'warning' wheel

Plate 72   Rear view of the Ellicott movement (Plate 71). The unengraved bracket for chime train and going barrel suggest that chime has been added to a fusee hour-striking clock

is likely that the six-bell rack chime is an addition, for its plain rear bracket does not fit well with the nicely engraved back plate, and its large going barrel is clearly as far to the right as it will go, making the winding square asymmetrical with those of going and strike. Whether or not it has been added to, and whatever its date, it is highly unusual, though following the basic principles of rack chime. The workmanship throughout is of a high standard.

Chime let-off is by the cranked flirt seen with its tip on a lifting pin (on the minute wheel with quarter snail) and its other arm ready to raise the pin in the rackhook. Strictly speaking, perhaps, it is a lifting piece rather than a flirt, since it is not thrown beyond its point of rest by a spring; on the other hand, there is no warning. The double-toothed rack, common on the European Continent, is not unknown in British clocks, but very rare; one set of teeth is for the gathering pallet, the other for the rackhook. The vertical rack is pivoted at its lower end, and there is not the usual cranked rack tail but a sprung finger, mid-way up the rack, to engage with the quarter snail.

The chime locking and hour-warning connection are individual. The central end of the rack carries a hollow box in which slides a block lightly sprung to rest downwards; however, when the tail of the chime gathering pallet presses against it, it is pushed upwards and the pallet locks against it. The other end of the block has the sprung hour warning lever, pivoted centrally, resting on it so that the hour warning piece is pushed down, out of the path of the warning pin when the chime train is locked – just as is normally achieved by pressure of the pin on the end of a conventional chime rack which connects with the

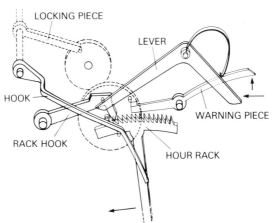

Fig 28   Detail of striking work by John Ellicott. Hour warning is held off by upward pressure of the quarter rack (not shown) on the tail of the warning piece. Strike is let off by fall of the quarter rack onto the lever, which causes the rackhook to rise. When the hour rack is thus freed, the hook falls, releasing the locking piece from the wheel next to the fly

120

Plate 73   British rack chime with warning and internal hour warning, c1830. Note left-hand lifting piece releasing chime rackhook, and also chime change 'pump'

Plate 74   View of the British rack chime (Plate 73), showing long heavy hammer for gong, its springing, the typical chime hammer array, and internal hour warning piece below pallet arbor on front plate

hour warning lever. When the chime rack is released at a quarter, therefore, the hour warning piece as usual comes into 'warn' position.

The hour striking is also unconventional, though I have seen parallels in longcase movements. The hour rack is released in the usual way by knocking out the rackhook when the chime rack falls at the fourth quarter, although in this case this is done by an intermediate pivoted lever working on a pin on the rackhook. The train is not locked by its gathering pallet, which has no tail and is behind the rackhook, but by the intersection of a large freely pivoted hook and a long pin on the rack. This hook is on the squared end of an arbor carrying a locking piece acting on a locking wheel with a pin (Fig 28). When the gathered rack holds the hook forward (as shown here), the locking piece is raised into the path of the locking-wheel pin. The locking wheel is in the position of warning wheel and the train warns on what would normally be the locking wheel – anticipating a common modern practice.

The chime hammers are pivoted in the usual slotted block behind a cover-plate, which also conceals the going-train fusee and the internal chime barrel. There was no room for the power supply between the

121

plates and so the large going barrel and intermediate wheel of the train were mounted on the back plate. It is difficult to judge when this movement was converted to chiming; for converted it almost certainly must have been. We may take it, I think, that the original movement, with at one time a verge escapement, started life as an hour striker, with the same locking system – which there was no reason to change – but also some additional striking system, for the cut-out to the left of the back plate appears to be original and is plainly now not needed. The addition of rack chiming may have been made early in the nineteenth century, and the design of the rack seems to show Continental influence. Whether or not the basic movement was by the great John Ellicott or another of the same family, it is difficult to say; certainly the compactness of the additional chiming system shows an ingenuity which he would have appreciated.

The movement in Plates 73–4 is interesting in relation both to the standard systems of rack chime with warning (Fig 27, Plate 70) and to later developments. It dates from the early years of the coiled-tape gong (1830–40), has a massive hammer and spring-stop arrangement which should be compared with that in Plate 27, on a smaller scale, for a back-plate bell, and offers the option of Westminster (then Cambridge) chimes and a form of Whittington chime on eight bells, the arrangement of whose hammers and springs is typical of the eighteenth century on. The selection is by means of a lever 'pumping' the internal barrel backwards against a spring on the back plate to give the alternative chime. Although the barrel is internal, it has in fact been made adjustable after assembly; its attached wheel, which with the barrel revolves twice an hour, can be loosened by a set-screw so that the barrel can be adjusted up to half a sequence. This of course enables the hammers to be set off the pins when the barrel is at rest. A dot on the barrel indicates the start of the first sequence.

However, the real curiosities of this movement are in the hour warning and let-off arrangements and in the fact that a large left-hand mounted lifting piece has been employed operating from pins in the cannon pinion rather than, as is more usual, in the minute and quarter-snail wheel. The other branch of this lifting piece is, as traditionally, the warning piece – passing through the plate, for the chime, but bearing a pin on an extension which raises the chime rackhook to let off chiming. Besides a forward-mounted pin for chime pallet locking, the quarter rack bears, rearwards, a long pin passing through the front plate and engaging with a forked pivoted lever (Fig 27a), the other end of which is the strike warning piece. This lever moves and effects strike warning with the quarter rack at each quarter, thus avoiding the somewhat temperamental usual arrangement where the adjustable warning lever is on the front plate, and also

Plate 75   Typical Edwardian rack-chiming longcase movement with flirt release. Flirt release and hook to silence; internal chime barrel for eight bells, external barrel for Westminster on gongs (pumping lever). Repeat lever (left to right across movement) knocks out quarter rackhook; standard vertical banks of hammer springs

Plate 76   View of the Edwardian longcase movement (Plate 75). Note flies with blades adjustable to vary speed; the substantial build of the movement; the long external spring for the heavy hour hammer, which is banked by a stiff upright spring

Plate 77 Edwardian rack-chiming bracket movement with flirt release, showing close resemblance to movement in Plate 75

Plate 78 Oblique view of Edwardian bracket movement (Plate 77), showing single pumped barrel to give choice of chimes on gongs only. Even in a clock of this quality, a massive going barrel has now ousted what once would have been a fusee

held out against a spring, by the quarter-rack pin. The present pivoted lever is unsprung, and always in engagement with the rack by pin and slot. Finally, hour let-off entirely avoids the normal knocking-out of the hour rackhook by the quarter rack, which is often a source of trouble. Instead, an eccentric hour lifting pin is used, in anticipation of modern and, as already mentioned, following Continental, practice. This raises the lifting piece further than usual, to such an extent that its pin raises the chime rackhook far enough to lift the hour rackhook and set the hour for let-off when the warning piece falls back into place.

Despite the recommendations of Britten, in the late nineteenth and early twentieth centuries when the countwheel supplanted the rack as the controller of chiming in cheaper clocks, rack chiming continued in more expensive pieces, nearly always with flirt release. This can be seen in Plates 75–8. Plates 75–6 show the movement of a very large and sumptuous longcase clock dating from the turn of the century. It is provided with chime/silent and tune-change levers in the break-arch of a reproduction brass dial with silvered ring and spandrels, and with a repetition lever operated by a cord within the case. Whether this was to show off the chimes or to repeat the dinner signal can only be guessed at, though plainly it was not for sounding the time at night. Both strike and chime have large external flies and the hammers are substantial; the sounding at both quarter and hour is very slow and impressive, on massive gongs for Westminster, and on an octave of bells for Whittington – a popular arrangement at the time. Overall, the construction is solid and substantial, with thick wheels, square edges to the cocks and brackets, and polished steel levers. The seatboard, appropriately, is nearly 5cm (2in) thick. The rack-chiming system was largely ousted by the countwheel method; but it died, and has been revived in reproductions, with imperial splendour. Such magnificent pieces, alas, do not fit into many modern homes, and the prices they realise at auction do not reflect their intrinsic merits.

Plates 77–8 show a movement of similar style and character in a bracket clock, although its weight determines that a bracket is not the best place for it to occupy. Here again are the break-arch indicators, the external flies with slow sounding on large coiled gongs, the high finish both inside and out and the same unsparing use of materials. The mechanisms are very similar. The flirt which lets off the chime is the shark-shaped piece pivoted in the middle of the front plate. Its small tail is depressed by the minute-wheel lifting pins acting against a spring so that, when released, the flirt flies up and knocks out the pin or projection on the chime rackhook. As Britten says, the energy taken from the going train to tension the flirt spring is considerable. There is no warning; the quarter rack is stiffly sprung to ensure that its tail falls fully onto the quarter snail before the quite slow-moving pallet starts to gather. As usual, the strike warning is controlled by a pin behind the quarter rack and is set by a spring when this rack is released. Again, the hour rack is released, with its pallet locking, by the fall of the quarter rack at the fourth quarter. A starwheel snail is fitted for the hour in both the longcase and the bracket movements, although there is no repeat facility in the bracket clock. The change-of-tune is by the traditional friction-tight lever with angled tail which pumps the barrel backwards; in the longcase movement the chime hammers are then out of line with their barrel and the bells cannot sound; in the bracket clock, a different set of pins on the barrel is presented to the same hammers. Silencing is arranged by two different ways of holding the flirt down against a lever – in other movements the flirt is pumped away from the lifting pins. The longcase movement's repeat lever raises the same pin on the chime rackhook as is struck by the flirt; this is not entirely satisfactory, since the rackhook may be released before the rack has fallen, or may be held up after chiming has begun, according to how the cord is pulled. The bracket clock has the additional luxury of a third indicator in the break-arch to control the rise-and-fall pendulum regulation.

To sum up, the superseding of rack chiming, even before self-correction was generally in use, is not hard to explain. The countwheel system is so much cheaper to produce and relatively little affected by wear that economic arguments outweigh the one intrinsic advantage of the rack system, namely that it cannot go out of sequence and presents less complication in moving the hands forward. Moreover, the rack action's great usefulness in hour repeating does not alter the fact that a chime barrel will continue on its way, so that the possibilities for true chiming repetition are very limited. However, the business of mass-producing chiming clocks is no longer what it was – the main-line timekeeper is not mechanical, and such a sounding clock is bought for pleasure, whether aesthetic or analytical. A three-train rack movement is a delightful mechanism and frequently has had devoted craftsmanship lavished on its finish. It sends us back nostalgically, if erroneously, to an age of luxury and leisure, and must be an enjoyable and increasingly valuable possession.

7

REPAIR OF INDIVIDUAL PARTS

This chapter lists principal parts alphabetically with notes on their adjustment, repair and replacement. The majority of replacement parts have to be specially made; hence there is no substitute for a wide knowledge of the systems and their working, and for reference to books containing plenty of photographs. However, some parts are often available in rough, or sometimes finished, form. These include bells, bell standards, cuckoo birds and pipes, gathering pallets (blanks), gongs, hammers, hammer springs, racks (blanks) and rack springs. On page 187 there is a list of suppliers who may be able to help, but it must be remembered that the best replacement available will generally be no more than a basic equivalent and a good deal of work will have to be put in before it can be used in your particular clock.

Bells and Bell Standards

Small bells are usually made of brass and larger ones, such as those on longcase clocks, of bell metal – an alloy of copper and tin, the exact proportion of each depending on intended use and the foundry. Glass was occasionally used in skeleton clocks where it was struck by wooden disc-hammers on wires; but this was far commoner on the European Continent. Until perhaps the middle of the nineteenth century, bells were cast; subsequently, they might be spun on a lathe. Cast bells are thicker and heavier for their size and much superior in tone. Spun brass bells, however, only dent if dropped and can be beaten out with fair success; cast bell-metal bells crack or break and soldered repairs are never very successful. Occasionally one meets an apparently cast bell which does not ring either truly or with a prolonged note; usually it will be found that the mounting hole is not in the true centre and nothing can be done. Where a bell is cracked but not actually broken, it will sometimes produce a reasonable tone if the crack is widened and filled with hard solder.

The fundamental pitch of a bell varies according to both the thickness of the metal and the diameter; bells of a given thickness sound an octave higher if their diameters are related 1:2. It is quite possible to tune bells at home on the lathe within certain limits; the difficulty is that tuning, either flat or sharp, involves removing some metal and this will eventually impair tone. To flatten a bell, metal is turned off the inside, so increasing the effective diameter; to sharpen a bell, metal is taken off the outside. This can be done using a tuned piano as standard and will produce acceptable results, for example in an octave of bells, provided you work slowly and listen carefully; but the dissimilarities between bell tone and string tone from a piano due to their harmonic structure are very great, and it is all too easy to think that you have hit the right note and to find next morning that you have cut too far. Note also that the heat generated in turning a bell modifies its pitch; let it cool before testing.

Tuning is a complex subject and the 'equal temperament' scales now universally used cannot be regarded as correct before about 1900. Before then, bells would have been tuned to some form of 'mean tone' temperament, and there seem to have been local variations. Therefore, if you are replacing a nest on an older clock and wish to be fairly correct, you should consult a manual on tuning – piano-tuning will do. Similarly, you should certainly do research before deciding that an old nest of bells is not in tune. In practice this applies principally to scales; the intervals are probably as near correct as you could come by modifying them.

The sound of a bell varies greatly according to its mass and the nature of what strikes it (see Hammers, page 146), but not much as to whether it is struck from the inside or the outside, or by a vertical or horizontal hammer. Lantern clocks and thirty-hour clocks, especially if they have posted movements, usually employ hammers striking inside whilst later and most eight-day clocks strike on the outside. Horizontal striking (Plates 31–2) is rare and early in British clocks, being far more general on the European Continent. What is far more important is where, on its surface, the bell is struck. The best tone comes from striking the thickest part of the edge, the bow, in a cast bell, and simply the edge in a spun bell.

Just as a tuning fork is virtually inaudible until touched onto a larger body which can amplify its vibrations, so the sound of clock bells cannot be divorced from their mounting and the structure beneath. The small vibrations of bells are almost lost in wood, and therefore bells have almost always been mounted on the movement rather than on the case; with gongs, save for carriage clocks, which are a special case, the opposite is true. Lantern clocks generally have their bells mounted in the cradle of straps at the top. Posted movements use a substantial support screwed into the top plate. Plated movements of necessity usually have the bells mounted on the plates – in longcase clocks on the front plate if early (Plates 28–30 show a later example), but most

128

generally on the back plate. Since the bell is the last piece in assembly, it often has to be removed when adjustment to the clock is needed. The odd convention by which the bell standard often comes to be mounted behind or through the pallet bridge (Plate 10) has already been mentioned; this is often inconvenient and there is generally room for it between the bridge and the hole for the strike fly. When high, hollow tops ceased to be the fashion – which they never entirely did either in bracket or in longcase clocks – bells were mounted on the back plate (Plate 27), often making access to, and regulation of, the pendulum awkward. This position was adopted in the small round French movements (Plate 15) very much earlier than in the traditional square or tapered heavy plates of the traditional British longcase clock or fusee bracket clock.

Nests of bells are found spaced with wood, leather and metal; but hardwood spacers seem to produce the best sound. The hammers of a clock chiming on bells are normally very light in weight and very lightly sprung, and they often rely on the flexing of a short shank to produce a rapid single blow. Thus the hammer heads at rest are very close to the bells, whose position calls for fine adjustment and firmer fixing than they often in fact received. Practice varies as to whether a nest is supported at both ends, but this is clearly a worthwhile measure.

Bells were to some extent superseded by gongs in the latter half of the nineteenth century. Early in the century coiled gongs were difficult to produce and more expensive than cast bells, whereas by the end of the century the opposite situation pertained. To a degree, that is still the case. Replacing good bells, particularly for chiming, is more difficult than replacing gongs where, save in carriage clocks, the size is less critical. Longcase and bracket-clock top hour-bells are readily available from suppliers in a good range of sizes. So are plated brass French bells. But the large flat hour-bells on the back plates of bracket clocks, and nests of bells generally, are a problem and the suppliers of new parts cannot help. The only course is to try stockists of old parts or to approach a foundry or supplier of clock kits containing such items (see page 187).

Quite often, particularly with top hour-bells, the bell and bell standard are missing and the bell is obtainable but the standard is not. Standards in thirty-hour clocks are usually unfinished, but finishing those intended for other clocks is a matter of graft with buffing stick and burnisher, and these standards are not difficult to make. You can make them in one piece, or you can fix bottom foot and top screw with silver solder. It is better to incorporate at least the foot in one piece – although this complicates things with a top-mounted thirty-hour standard with the foot at right-angles – by hammering out roughly and bending while the steel is red-hot for softening. The cooled metal can then be filed or ground down to a taper and bent round at the top. Practice seems to have varied as to what sort of stop or seating, if any, was provided for the bell to rest upon. It is reasonable, if untraditional, to flatten the top of the standard, drill it, and fit a bolt and nut, so that the nut will act as stop. More proper practice is to ensure that the top, when bent, is thick enough to be reduced and tapped, leaving a stub for the seating. Occasionally bell and stub have a square hole.

Steady pins are sometimes used in the feet, but more often the tip is bent back to lodge in a hole in the plate. In my experience the cast bells from suppliers have holes that are too small for old bell standards. I mention this because, since holes can obviously be broached out, it is a mistake to tap your standard to fit these holes, and much better to use, say, a 2BA screw thread and an enlarged hole.

The usual brass standard on the back plates of French clocks is occasionally available from suppliers. It is not easy to make, because of its relatively large area and the thickness of the foot. The best course seems to be to make the foot separately and then to braze or silver-solder it to the stem, which can be bent out of brass rod. The top end is bent to a right-angle after heating, the bend being hammered sharp, and is then tapped and fitted with a brass bush for the bell seating.

Chime Barrels

Barrels and their equivalents exist in many forms. We have noticed barrels of the 'musical box' variety – the traditional sort (Plates 42 and 73), toothed discs clamped together (Plate 62), pinned discs screwed together (Plate 57), and 'snagged' steel cylinders (Plate 60). The snagged variety is not easy to repair – it is seldom damaged, but it does wear and the teeth can be raised and re-shaped. However, the wear on the hammer nibs is likely to be worse, though they can be refaced. Pins in the pinned-disc variety can be replaced easily enough, provided that the discs are clearly marked for reassembly in the correct positions. These work on the pin-wheel principle of horizontal pins, but in all other cases it is advisable to test worn and repaired barrels in the lathe since it is essential (except in some rare old barrels, see page 133) that all pins and teeth be of the same height. Nor is the replacement of the toothed-disc type a great problem; there are many around to be had for a song still in dilapidated movements and often of the correct size. But, unless you have an exceptional barrel in a valuable antique, it is best to think in terms of making new barrels of the cylinder type when you are confronted by seriously damaged or missing chime barrels. These are not difficult to make or repair.

There seems to be no rule governing the size of barrel appropriate to certain trains or power supplies. Generally this is a matter to be worked out empirically when making a new chiming movement rather than repairing; on an old movement the size should be clear within reasonable limits from the position of the hammer-arbor hole (or the hammers if they are there) and the barrel-arbor hole. The length and distance between pins will also be obvious if the hammers are there and, if not, you can please yourself subject to the need for an internal barrel and its wheel – which you will have to calculate from the driving or ratio wheel – to fit between the plates and any separate bracket there may be.

Barrels are usually made from brass tube, which needs to be at least 2mm ($\frac{1}{12}$in) thick in the wall, but copper is also satisfactory provided you work it at a high speed in the lathe. Cut and face the ends to form a cylinder of the required length. Turn two blanks of about 3mm ($\frac{1}{8}$in) or more in thickness and drill them to be a close fit for a pivoted arbor of the size required. Tin the edges of the blanks and of the cylinder with soft solder; the blanks should be a free fit to allow room for this to be done. Insert one blank, stand the cylinder over it and heat until the solder flows. When the cylinder is cool, repeat the process, with a solid block of metal on the already soldered end to keep that part cool. The arbor can then be fitted and either soldered in or fixed with Loctite 601. Set up the barrel and arbor in a chuck and, with a centre to support the other end, turn it true; afterwards polish to the desired finish with an emery buff.

At this stage you have to decide whether to mark the metal or to make a paper pattern, marked up, to stick round the barrel showing where the pins are to go. Either way, the barrel must be indexed to the total number of divisions in the chime (see below), and then lines are drawn right round it to mark the position for each hammer, not forgetting that if there are repeated notes in the chime, they will probably require two hammers. You then have to drill the barrel for the pins either from your separate pattern, or from the pattern already stuck or marked on. The Unimat vertical-column drill assembly is ideal for this job, but it can be done with any vertical drill or even with a hand-drill. The drill chosen must make a hole which will not quite admit the size of pin chosen – for this purpose decapitated panel pins seem quite adequate – and life will be made easier if you can use a centre drill and avoid having to punch the mounted barrel.

When fitting the pins, support the barrel wherever possible beneath the workpoint; this of course becomes more difficult as you go and fitted pins begin to obstruct. Therefore, when making a new barrel or repairing an old one, make a hollow punch, the hole in the end nicely fitting your size of pin and exactly the depth which will give its required height. Subject to slight irregularities where they have been cut, your pins will then all be of the same height. If the metal is sufficiently thick and the holes nicely judged, the pins will be firm enough when driven in dry, but you can if necessary give each a small drop of Loctite to make sure. When the whole barrel is pinned, the index can be removed and the pins must be dressed to size. This is easily done with a slip-stone in the toolrest, revolving the barrel at high speed, although it takes some time. Finally, the ends of the pins should be rounded using a home-made chamfer (a cross, filed with a triangular file on the end of silver-steel rod) or a commercial jeweller's cup-fraise.

In all this, while a hole in the wrong place is not a disaster and pins can be extracted and moved later if the timing of the chime is not quite right, it is vital to have a very clear idea of what you are about and to get the right note in the right place and the spaces as intended. Whether you mark the metal or use a paper pattern, it is therefore wise to make a 'map' before you start, and it makes sense to mark on the barrel the first sequence. A map for Westminster chimes is shown in

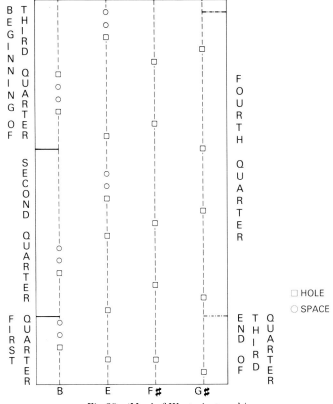

Fig 29 'Map' of Westminster chimes

132

Fig 29. First make columns, one for each note in your chime, on squared paper. Take the smallest musical unit in the chant – usually one beat, a crotchet – as representing one square, and count how many of these units there are altogether. For a double-length note, you naturally allow two units; for the space between each sequence you need at least two units. If you are using a conventional chime, there will be no difficulty in that each sequence will contain the same number of units; but if, for whatever reason, you are inventing a chime or using a 'non-chime' tune, it is essential that you convert it into five sequences, each with the same number of units, though adding the odd space here and there to meet this criterion will not show. Normally, these five sequences will go round your barrel. Exceptionally – and a look at the ratio and driving wheels will tell you – the barrel goes round only once in an hour, in which case your five sequences will have to fit onto it twice over. Once you have done the calculations, plot the map, using a cross for a pin and a circle for a space, or some similar arrangement, the pins of course being in the appropriate hammer columns. As a safeguard, do the drilling with the map beside you, ticking off pins and spaces as you go; it is all too easy to be distracted and to miscount in this job, and if you get out of step, rather than misplacing the odd pin, you will have to start again with a new cylinder. If there is more than one tune on the barrel, you need two maps, and it is wise to pin one tune completely before starting on the next.

You may find, when the barrel is fitted, that it is impossible to stop the chime without leaving the hammer on a pin. This indicates either that the barrel is too small or that the pins are too long. When replacing pins on an old barrel, watch out for varying heights; these were occasionally used to give varied dynamics, and your new pins should at least sound sensible.

If you want practice in making barrels, particularly to novel tunes, an interesting and rewarding exercise is to convert a cheap Westminster chimer from the 1930s to sound on eight gongs. The chief problem is power; it is advisable to fit a new mainspring – the original one will be exhausted anyway – and it may be necessary to shorten and lighten the existing hammers, besides adding four (or three) of similar pattern. The present train will almost certainly run too fast, so that it will probably be necessary to make a new fly, larger but light, outside the back plate. It is simplest to follow what was once a common practice and make your hammer and barrel assembly between two sub-plates screwed to the back plate as a detachable unit. The barrel will, of course, extend further back so that you will need to extend the case or make a new one. The diameter is limited by the power available, and about 4cm (1.6in) seems to be the maximum possible. This means that you cannot use a tune or chime having more than about seventy units, but that will accommodate most songs and hymns in common metre, with a fifth line added, or an earlier line repeated, as the fifth sequence. Considerable juggling will be required to match load and power available, but you will learn a great deal about barrel making and you will probably give an interesting new life to a movement destined for the scrap-heap.

One can be presented with a clock complete with barrel but lacking gongs or bells, and the question then arises as to what notes are involved and whether the pitch is highest at the back or at the front. It is not too bad with a relatively modern clock, for the chime is likely to be a recognisable one and may well, as many do, start with a downward flight or scale which will be apparent as a diagonal line of pins or teeth on the barrel. Again, with a clock from the 1920s onwards, and even before, it is more than likely that 'Westminster' is one of several chimes. But with older clocks it is not so easy. Clearly a repeatable chime is recognisable by repeated patterns of pins, usually downward scales; but if the pattern is more complex the best approach is to imagine yourself designing the barrel and make a 'map' of the pins, numbering the columns. You can then convert this map onto a musical stave using any major musical scale as basis, and testing either end of the barrel as top or bottom note. If this fails to produce a shapely tune, you will have to try with other notes in the scale as top or bottom; this will, of course, change the place of the two semitones found in the major scale. If there are eight bells, you will by then probably have found the solution, because the eight are likely to be an octave, and even six bells are usually adjacent notes. Where, however, the intervals are larger than tones and semitones – as they will be with some six-bell and most four-bell chimes – you will have to experiment further. You may or may not eventually arrive at one of the listed chimes in, for example, De Carle's *Watch and Clock Encyclopedia* or Jendritzki's *Repairing Antique Pendulum Clocks*, and even after study if possible of chimes by the same maker you may have much doubt as to whether you have identified the chime, but you should have an acceptable tune. Although the clock by Ellicott in Plates 71–2 came to me complete and presumably correct, I have never been able to identify the six-bell chime and suspect that it may be as unusual as the chiming system itself.

Countwheels

Countwheels are among the most frequently missing parts of striking and chiming clocks, since so often they are external and held on by only a pin or a spring clip or washer. Fortunately, they are not difficult to replace. The question of repair does not, or should not, arise, unless the wheel has been tampered with in an effort to correct what was in fact some other malfunction.

The immediate question that arises is one of size, and this is best discovered from observation of the striking or chiming train. Indeed, the whole job can be done very satisfactorily and reliably by markings on a blank drilled and held in place. The crucial point to watch, whether in striking or in chiming, is the locking; for the countwheel system entirely depends on the interaction of locking and countwheel. The locking is likely to be either some form of hoopwheel or cam, or a locking piece butting on a wheel pin. If this is missing, you must make it, and the same applies to the countwheel detent (see next section), modifying as necessary as work on the countwheel proceeds. The depth of the countwheel slot must be such that, when the train is to be locked, the locking piece can reach the bottom of cam or hoopwheel, or the wheel pin is well engaged with the locking piece; and, when it is unlocked, the locking is correspondingly completely clear. With British external countwheels this allows a good deal of freedom, and the slots are often deeper than they need to be and the detent rises well above the wheel. With French countwheels, on the other hand, the locking piece and countwheel detent move very little, the slots are seldom deeper than they need be and the detent in fact rides on the countwheel edge when it is raised. The pin-countwheel presents the slightly different problem of at what radius the pins should be set, bearing in mind that the hooked detent passes above and below them. In fact, there is a pin, or its hole, placed as a stop for the detent; and the detent must not be allowed to fall below the point where pressure from a countwheel pin on its sloped face will push it up. This roughly determines the lowest point, and after that the detent can be bent to the rather critical point where the system acts reliably. If you have no guide from a previous countwheel, do not make the pins too thick, as this can cause double locking. Thus, as a rule, the size of a countwheel is best found out on the clock, and here a cardboard blank comes in useful. Try it with various depths of slot, and cut it down until the action is satisfactory. All this is best done in conjunction with the associated gearwheel which, in a British clock, will be missing – as probably will be its pinion. You should be able to calculate these from the information in Chapter 2 and either make them or have them made, for example by an advertiser in clock magazines.

There is no need for strength in a British countwheel, but one made of thin sheet brass is distasteful; it is much better to use a piercing-saw on metal over 1mm (30 mil) thick. These wheels are never crossed out and the work is straightforward enough. The wheel is based on 78 divisions, and, as 78 is not a factor of 360, it is easiest to measure round

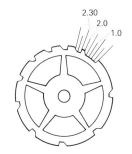

FRENCH COUNTWHEEL
(90 DIVISIONS OF 4° EACH)

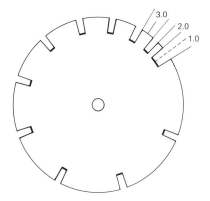

BRITISH COUNTWHEEL
(78 DIVISIONS OF CIRCUMFERENCE)

Fig 30   British and French countwheels

the circumference (πr^2) equal divisions (Fig 30). The exact size is not critical; choose a circumference which is 78 times a particular unit such as .5cm (⅕in). The result may well not be accurate at every hour if you do it this way and you are sure to have to seek out the best position for the wheel. Therefore the simplest and safest way is to make up the gears, attach the wheel loosely to the countwheel, try it with the train running and mark wherever it stops at each count – the count of one will give the width of each slot.

The French countwheel can be made from similar material and by similar methods. It fits on a square and there is no complication of lost gears. It must, however, be more highly finished and crossed out, and there should be a domed collet between wheel and securing pin. There is, customarily, a slight groove between inner and outer edges and this, though not essential to working, requires a lathe. The wheel is based on 90 divisions (Fig 30), and as 90 is a factor of 360 there is a good chance of sufficiently accurate sections by dividing the circle. Because the

136

French movement locks with locking piece and pin, ie there is no slope on hoopwheel or cam to raise the countwheel detent, the requisite lift is given by slanting the oncoming edge of each countwheel section. When dividing the wheel, make the base of each slot 1½ divisions, so that at the top of the slope the next section begins on the second division.

The basic chime countwheel is made round 10 divisions (Fig 31), that being the number of sequences in an hour, and it may be constructed of brass or steel. Beyond the necessity to accommodate the locking, as with hour countwheels, the fourth section has usually to be raised so that the strike flirt knocks out the hour rackhook. Try it with a cardboard blank. Older countwheels of course use a pin for this purpose and the shape and angle of their slots will be determined by the shape of the countwheel detent (Plate 52). The chime countwheels of the 1930s seem to have rounded slots even if the locking system in fact will raise the detent. The oldest chime countwheels were circular, with slots like those of British countwheel striking, but they are very rare pieces and one would be well advised to seek specialist advice in their restoration.

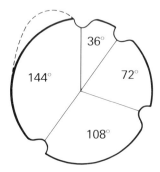

Fig 31   Chime countwheel

The projecting-section type of countwheel, found most often on Black Forest and cuckoo clocks, seems sometimes to have been turned from solid (Plate 21). Normally, however, this is hardly justified, and the countwheel can be made by cutting – or having cut – its geared base, soldering round the edge a rim made of tubing of the required diameter, and then cutting out the sections. It is wise to mark these on the wheel, or a paper pattern stuck to it, beforehand.

Naturally, if you have a lathe or wheel-cutter with dividing plates, the construction of any countwheel is more accurate and much simplified; but you are unlikely to be involved in batch production and it is quite satisfactory to cut out the slots with a piercing-saw rather than make a fly-cutter for the purpose.

Countwheel Detents and Locking Pieces

There is no difficulty in making these pieces from strip brass or, more often, steel, but their acting surfaces should be polished and the angles of engagement are very important. It will normally be apparent from the countwheel, locking wheel, cam or hoopwheel that a lifting action is required. The thirty-hour hoopwheel locking piece has a dead face for locking and a curved face for lifting (Fig 32). The French detent has a rounded profile to match the lifting slope of the countwheel slots (Plate 15). The hooked detent of the pin-countwheel (Plate 15) has to be tightly bent, filed off and shortened so that it will rise when struck by a countwheel pin, rest on the pin, then fall right off it; if the hook is left too long, the train will lock again after striking only one blow, as the detent fails to fall off the pin.

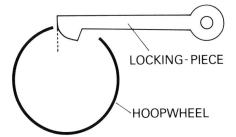

Fig 32   Vertical locking of hoopwheel locking piece

Generally, bladed detents, such as on old British and American movements, should enter the countwheel slots as nearly as possible radially to the centre of the wheel. This indicates the length of the detent as well as the angle of the blade. If there are hour slots only, the blade should enter the centre of the slot. If there are half-hour slots, the blade should enter at the beginning or there will not be room for another entry at the half. If the detent on a countwheel with straight slots enters too late, there is a risk that it will butt on the exit wall and jam the train at warning; the countwheel plays no part in the actual locking and the sides of the slots should not be touched by the detent. The chime countwheel detent should rest at the bottom of what in modern clocks is usually a rounded hollow; with older chime count-wheels the slots may be angular (Plate 51). The locking action will not be reliable unless the detent falls in such a way as to fit the slot.

Locking angles are even more critical than those of countwheel detents. With hoopwheels and cams, particularly if the locking piece catches from behind, the faces must be vertical and the locking face may even be slightly undercut (Fig 33). Similarly, when the vertical face is met head-on, shaping the piece back to give a little draw is

138

Fig 33   Locking with 'draw' on a cam

Fig 34   Locking of different British pallet tails

beneficial; if it is even slightly less than vertical in the other direction, mislocking is likely to occur. Pallet locking on the traditional British rack demands both that the train locks reliably and that it can be released with the minimum of friction; gathering-pallet tails have either two angled faces or are semicircular (Fig 34). Locking should occur on the edge of the 'flat' surface, and draw is not required to assure locking if the rackhook is engaging properly without play on the first tooth. With the modern rackhook, locked by the cam of the pallet (Plate 46), the projection of the rackhook – which does the locking – may have to be twisted slightly according to the shape of the pallet cam, until engagement between the two is firm. The locking piece and rackhook of Comtoises can be found bent relative to each other; they must be adjusted until the train is firmly locked when the hook is underneath the raised rack, but free when the hook is in the teeth during gathering (Fig 35). The same applies to the usual form of French rack striking (Plate 33), where again the locking face must not be angled away from the locking pin in either direction.

As the locking piece for the hoopwheel of a thirty-hour clock has two faces, one vertical for locking and one sloped for lifting itself and the countwheel detent, it is essential that the piece enters the hoop slot far enough to lock the train beyond all doubt and rebound. What will usually prevent it from doing so is a bent countwheel detent or a new countwheel with too shallow a slot, but it may be necessary to file the vertical face straight also.

Particularly with the pin-countwheel type of movement, but also with any of the many types (including nearly all French clocks) which employ a locking piece which obstructs a pin in the locking wheel, wear, or indeed bending, can alter the point on the surface at which the pin strikes. The engagement must obviously be deep enough to lock positively, but not so deep that the pin rubs the locking piece when the

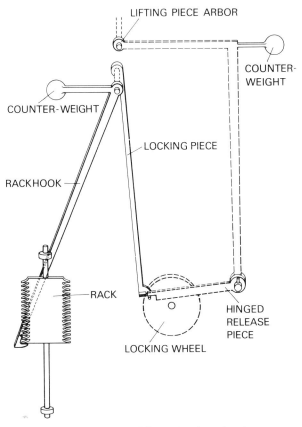

Fig 35   Locking of Comtoise (from front)

train is running. This appears straightforward but can be quite difficult to obtain. The cause of the trouble is nearly always not the locking piece itself but the rackhook or any other piece which fits on a squared extension of the locking arbor. If such a piece itself is bent or if, as so often happens, it has some play on the arbor, consistent locking will be very difficult to obtain. On modern chiming clocks, where this arrangement may also be used for warning, the locking piece is nearly always adjustable by set-screw, which greatly simplifies matters.

Obvious though it may be, it is the cause of much trouble both in counting and locking if a detent or rackhook is pinned loosely onto a squared arbor. It must be absolutely firm on the arbor, for a sloppy fit upsets the geometry on which the whole mechanism depends. If the piece has simply too big a hole, this can sometimes be closed with a hollow punch or wedged with a slip of metal. It may also be possible to stretch the square by hammering – but such defects have a habit of not becoming evident until after assembly.

140

Flies

Although the fly has a fundamental effect on the speed of sounding, various escapements and similar devices have been proposed as alternatives; and the gearing is quite as important as the design of the fly. Usually, as we have seen, a small fly pinion is used (generally of six leaves, occasionally even of five), and the ratio to the warning wheel has tended to increase. Flies have become larger and lighter to act as air-brakes and, for the purpose of slowing the train, that is fine; however, the fly also has a role in evening out the strains of hammer pick-up and release, and for this purpose a certain mass is probably desirable. The heavy hammers and strong hammer springs of early clocks are to some extent a justification of the, by modern standards, heavy flies employed.

Seventeenth- and eighteenth-century flies are mainly of standard two-bladed type, and relatively small and heavy. Sometimes the restriction of their being between the plates led to making cut-outs in them to clear a turned pillar or the pallet anchor (Plate 12). Later, it was common for lighter flies, still between plates, to be cut out to clear the warning wheel. All these simple flies can be fairly simply made in two basic ways: either the shaped sheet brass is dented for the arbor with blows from a wedge-shaped piece of hardwood, and the riveted brass strip spring holds it in place; or the arbor can have proper round holes, the fly being milled down from thicker metal or blades being hard-soldered into brass bushes to fit the arbor. The brass spring should be hammered to harden it and, although I have seen apparently original springs soldered on, it is usually riveted with two brass pins. French-clock and simple modern-chime flies are similar, but the spring should be steel – a piece of watch mainspring or thick shim – bent into place as shown in Fig 36.

In the nineteenth century more elaborate flies became common (see Plates 57, 73 and 76). The centrifugal type is designed to stabilise the

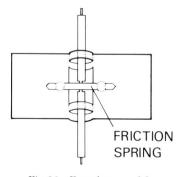

FRICTION
SPRING

Fig 36   French type of fly

141

speed, the fly offering more resistance and opening as the train gets under way. Sometimes one blade is fixed and has a small weight on it whilst the other is sprung and can swing outwards, and sometimes both blades can open. Particularly on more expensive clocks, flies are brought outside the plates with an extended arbor and separate cock. These have large light blades whose angles can often be varied to set the sounding at different speeds. The Comtoise always seems to have used the four-bladed large fly shown in Plate 32, which was cut away in the centre to accommodate a coiled spring working against the pinion face. Such a fly can be made by slotting blades into brass bushes and securing with Loctite.

It is essential that flies, of whatever type, be balanced, otherwise the ability of the train to overcome initial inertia will be seriously impaired. The balance can be tested by setting the pivots on the jaws of a suitably opened vice. Adjustments to the simpler flies can be made by filing a little off one blade. With others it may be preferable to add a small lead pip to the light side. The limits to the opening of centrifugal flies are usually set by the butting of the open blade on the fly's end-piece, but arrangements vary. These flies, particularly those where only one blade opens, must be set to a reasonable balance when closed, regardless of what may happen when the blade opens.

Flies may have adjustable bearings with the pivot hole at the front in a plug with an eccentric hole. The plug may be scored with a screwdriver slot, or have a raised section to turn with pliers. In movements after about World War I, the same end is achieved by cutting out this part of the plate so that the pivot hole is in a mere twig of metal, which can be bent. These bearings are adjusted for free and silent running – a somewhat risky procedure with fine pivots. Ideally, the depth should be tested before assembly since turning a tight bearing can easily break a pivot. A plug with a badly worn slot should be knocked out and replaced with a newly turned substitute. These bearings can be exceedingly tight and if all seems well they are best left alone.

Gathering Pallets

Blanks for British rack pallets are available from suppliers; sensibly, they are on the large side and considerable reduction will be needed in most cases both to tail and pallet. French, usually tailless pallets, are not so readily available and will have to be made up from steel stock. They are very small, so that it is best to file up the shape of the pallet and to drill it, on the small side, before separating from the parent steel. In all gathering pallets, the pallet tooth should be radial to the hole. When separated from the stock, the pallet hole will have to be squared or tapered (or both) with needle files and broaches – a slow job. The shaped pallet can be hardened (heated to bright red and plunged into salt water) and tempered (heated just beyond dark blue and cooled in oil), before being tested, finished to size and polished with emery buffs.

Modern cam and pin pallets can be filed from brass according to the locking requirement and according to whether or not they are required to move the rackhook aside as a tooth is gathered. The position of the pin – which should be of pivot steel – relative to this lifting action is critical, and will have to be found from trial as you lead the brass round against the rack. Usually the pin is on a radial line between the locking face of the cam and the arbor, and opposite the tip of the cam (Fig 37). The essential with these pallets, which do vary, is that where the rackhook does not have ratchet action but is controlled by the pallet cam, the rackhook must be allowed to catch a rack tooth just before the gathering pin releases the rack.

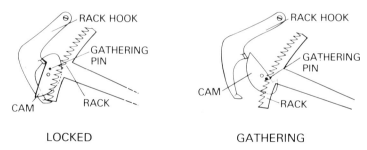

LOCKED   GATHERING

Fig 37   Pin and cam pallet action. When the gathering pin is fully meshed with the rack, the tip of the cam holds the rackhook clear of the rack; no ratchet action is therefore needed.

When making solid pallets, start with the tooth on the large side and only very gradually reduce it; for it is very hard to correct a short pallet. For safe action, the pallet must be long enough to raise the tooth so far that the rackhook has to drop back a little after it has been dragged over one tooth. There should never be occasion to shorten an existing pallet tooth, for very probably the rack circle is out of true if the pallet seems too long; but wear may make a pallet short. If so, a new pallet may not be justified, and a reasonable repair can be made with a slip of mainspring soldered, filed down and polished, as in refacing a recoil-escapement pallet. The other major cause of trouble with a gathering pallet is a worn hole for the arbor; a loose hole here must be properly bushed before any other adjustments to the gathering are attempted.

The shape of the pallet tail, if there is one, depends on the original if you have the pattern or, if you have not, on the position of the rack pin and the best compromise you can work out between locking and easy release (see Countwheel Detents and Locking Pieces above). Equally important is whether the tail is in front of or behind the rack and the rest of the pallet. If the pin and pallet are behind the rack you require a left-hand blank; and if they are in front you require a right-hand one if, as is most common, the rack is gathered from left to right. The tail of the blank may be filed into a curve, either semicircular or at an angle. The rack pin should come just at the start of the changed shape. On no account must it be on a reverse angle where the rack will have to move the train backwards to be released; this will necessitate a strong rack spring and impose quite unnecessary strain on pallet and rack if the train runs at all. The shape of the back of the tail is usually of no significance (see Plates 23, 27 and 43) where it releases the caught flirt), but the back of the tooth and the whole boss will need considerable reduction until they properly clear the rack teeth.

The gathering pallets of Comtoise and old British internal-rack clocks are distinctive in being part of the arbors; the British arbor is built up to a blade, and the Comtoise pallet is a single pinion leaf flattened. Replacement of the whole arbor for the sake of a worn pallet may no doubt be the ideal repair, but refacing with spring steel will last for some time and need not be unsightly.

The double and triple pallets found in quarter-striking and grande-sonnerie carriage clocks must be specially made for their racks, remembering that the back tooth must be long enough to raise the flirt and release it from the rackhook as the pallet revolves.

A rather time-consuming repair is sometimes required in English pallet systems, where the pallet and square can become twisted off, leaving a short arbor and no pallet. Unquestionably the best repair is then to make and fit a complete new arbor, filing the square to fit the pallet hole (where of course the pallet itself may be a replacement). Short of this, however, it can suffice to drill the shortened arbor to accept a new pallet-mounting square, which is turned down to fit the hole closely and fitted with Loctite 601 or, preferably, silver-soldered into the arbor. Fast-striking hour trains often result in considerable strain at this point – the square must be true, and there must be no chance of its turning on the arbor.

Gongs

Gongs are either coiled-tape or round-sectioned steel, bronze, or phosphor-bronze rods; or tubes, usually of brass. They have a large amplitude of vibration and have to be struck with a relatively soft-faced hammer if – to modern ears – disagreeable clanging harmonics are to be damped out. Gongs came to Britain from the European Continent,

144

where coiled ones were introduced in about 1820 and rods and tubes in about 1880. High-class clocks retained at least the option of chiming on bells, and Grimthorpe (1874) was not very struck by 'those spiral springs on which small clocks sometimes strike'. He was probably thinking of carriage clocks and small Black Forest clocks, which were among the earliest users of coiled gongs, although simple gongs had been used in repeating watches in the eighteenth century.

The difficulty with repairing gongs is that heat destroys their temper. A quick soft-soldering job is the best hope. Neither rods nor coils can be bent beyond their point of return or the tone is ruined, but bending in the straight stretch near mounting is possible and the very thin gongs such as are used in cuckoo clocks can be bent without serious harm. Rust spoils appearance and tone and should be removed with a very fine buff, polishing with hardwood. The gong cannot be reblued in a flame or it will be ruined, but the heat required for blueing-salts will not impair its tone. Because of their large vibrations, gongs require substantial mounting with washers to support the column. Except in carriage clocks, they are mounted on the wooden case; and the quality and size of the latter, acting as a soundboard, play a great part in their tone.

Good quality replacement coiled gongs are very hard to find, and dealers are the best hope, though the wire Black Forest gongs are available from suppliers. Tubular gongs, likewise, can be obtained only from dealers or by making them yourself, which is quite practicable although you need a large quantity of extra tube owing to the experimentation involved in tuning. It is wise to buy, for an octave, at least four lengths of your maximum size and another four lengths three-quarters of that size. Rod gongs are available from suppliers, both loose and in tuned sets.

It is quite feasible to tune rod and tubular gongs against a tuned piano. The secret is to go slowly and never to risk cutting off too much metal, or you will have to start again with the whole series at higher pitch. Start with whatever note you decide for the longest gong and proceed from there. With a rod gong tuning must be done in the support and it is essential that these gongs are screwed in as tightly as possible; older ones were driven in and sometimes held with a grubscrew. Whilst it is always possible to raise pitch by shortening a gong or tube, flattening is a different matter. If you do make a mistake and end up slightly sharp, a tube's tone is not greatly affected by soldering a small piece on, and a rod gong can be flattened by careful filing at the thinnest point by the screw. So far as I know, it is impossible to obtain coiled gongs with which to make up a set and so the question of tuning these does not arise. In all tuning operations, as with bells, slow, careful work is particularly necessary when you are approaching the desired pitch. The fundamentals of a piano and a gong are not easily compared; you must listen very carefully, testing also against octaves above and below, and removing very small amounts of metal at a time. With identical metals, a gong an octave apart is twice the length; a semitone is, near enough, one-twelfth of the difference in length.

There are two more, perhaps obvious, points to mention. The first is that thickness as well as length of gongs affects their tone and pitch. Therefore it is often not possible to replace one of a broken set, for example with a shorter but fatter one, and you may have to replace the whole set. Secondly, while it is probably desirable to strike a gong at one of its nodes (a factor of its length corresponding to its harmonics), in practice it is sufficient to ensure that it is struck near its mounting point where the amplitude of vibration will be very small, this being even more desirable if the gong is struck with a metal-faced hammer. Tubular gongs must be struck radially, ie as seen from above, or their tone is much impoverished.

Hammers

The length of shaft and the mass and composition of a clock hammer influence the sound produced. Just as over the past 150 years piano heads have moved from relatively hard materials to carefully compressed felts in the search for soft and mellow tones, so softer faces have been developed for clock hammers to suit vibrant gongs, rather than the brass or metal faces employed with bells. The softer surface lingers on the gong and damps out the strident sounds that a metal hammer produces; a bell's sound, on the other hand, is characterised by just these harmonics vibrating at lesser amplitude. It follows that the leather or composition head of a softened hammer requires adjustment to produce a pleasant sound consistent with the other gongs in the series. The amount of facing material can be reduced or the surface hardened by singeing with a soldering iron; alternatively, the leather may be oiled and teased with a needle to soften it.

No such adjustment is possible with metal hammer heads. These should have a rounded striking point and their mass should have a reasonable relationship to the mass of the object struck, being related also to the spring which impels them. The chiming hammers in Plates 51 and 75 are of streamlined brass about 3mm (⅛in) thick, whereas the hour hammer of any longcase clock tends to be a square steel block 12 to 20mm (½ to ¾in) square, shaped down to a rounded point. Such hammers are frequently broken off at their arbors and there is no difficulty in forming new hammers and shanks, filing a slot and brazing them in. The brass chiming heads often come loose – they are bent to adjust their angle to their bells – and this will impair the sound; the hole should be cleaned out and the hammer fixed either with solder or with Loctite.

Chime hammers, new and old, depend on freedom of mounting, but yet they must rise and fall truly without wobble. The old practice was to mount them in a slotted block, with a pin to hold them in place. This should always be dismantled and oiled. In cases of serious wear it may be necessary to make a new block, but if not all the hammers are affected it is possible to face the worst ones with shim steel. Where a new block is being made, there is a notorious difficulty, and standard solution, to making the fine hole for the pin through so long a piece of brass. Instead of drilling, saw a slot of thickness equal to the pin's diameter, and then fill this slot, save for the space for the pin at the bottom, with a slip of brass soft-soldered into place. The more modern practice is to mount the hammers on an arbor with spacers; and here wear can be taken up by adding a collar with set-screw, if one is not present, to hold the hammers together. The heels, or steel nibs, of chime hammers sometimes wear quite badly where they are scored by the barrel pins. The result of this is not normally any worse than a softening of the hammer blow. In bad cases bending the nibs or refacing the heels is not likely to be satisfactory for long, and it is better to make up new parts if the clock is worth it.

All hammers driven by a powerful spring, and this amounts to most in which the hour bell is above the clock, acquire serious wear in their arbor pivot holes. This impairs the action of the hammer and may make it difficult to set up off the pinwheel pins. The only solution is to bush the holes. Round French hammer heads usually have central holes, so replacements can be turned, ringed with a round tool, in the lathe.

In the same way as chime hammers twist on their shanks, so notoriously do the brass-wire shanks of French hammers twist on the square-holed bosses by which they are pinned to their arbors, and again the cause is the bending of the hammer to strike the bell in the right place. One comes across cases where the joint has clearly been soldered and soldered again to no effect. This is a case where modern science may well step in: clean out the hole, file away all the solder and insert the shank with a drop of Loctite 601 or super-glue. It will look better, be stronger and not be a hindrance to later repairers.

Finally, as an indication of what weird arrangements one can find for the apparently simple task of causing a hammer to strike a gong by means of a revolving wheel, look at the little Black Forest clock in Plate 22. Studying it, you will see that the hammer wire is pushed forward by the arbor against a vertical coiled spring and is curiously looped to make this possible. As a result, the hammer descends by the drive, but relies on the coiled spring's force against gravity to raise it to strike the gong.

147

Hammer Springs and Stops

For our purpose the earliest type of spring and stop is that illustrated in Plate 3, where the stop is a very stout iron 'L' terminating in a screw through the top plate and the spring an equally sturdy straight piece of iron, similarly fixed to the top plate and working on a stub raised from the hammer arbor. This is the usual arrangement for a posted movement with the hammer striking on the inside. The stop is not rigid; the hammer itself has no flexibility and relies on the spring in the stop-piece to make its blow only momentary. Both spring and stop must be firmly screwed up but can be bent to take up wear. They frequently show the marks of a forging hammer and replacements can be made by beating out pieces of iron and brazing on the screws. The spring should be tapered if it is to work effectively.

This vertical spring, taken over from lantern clocks, was simply adapted to movements with plates by changing the plane of the foot in relation to the spring, as can be seen in Plate 6 where the foot is screwed into the plate from inside, or Plate 10 where the screw goes into the foot from outside, the latter arrangement having the advantage that the spring can be inserted after assembly. Although the hammer is now outside the bell, the spring still works on the inside of the arbor because the relationship is now reversed and the spring comes from below. This type of spring continued in use for at least two centuries in fusee-bracket and some longcase clocks. It is less easy to make, because of the angle of the foot. Therefore whilst such springs were clearly forged in one piece it is more practicable for us to make them from two pieces – a short strip bent into a right-angle for the foot, and the long spring silver-soldered to the foot's raised edge (Fig 38). Again, you should try to give the spring a taper or a curve towards the hammer arbor, depending on space available.

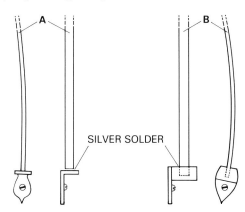

Fig 38   Making standard hammer springs (two different patterns)

148

All types of stops are found with this simple spring. The hammer shanks may become much thinner, use being made of their flexibility, and rigid stops may be employed against a stout pin in the arbor (Plate 7). Sometimes the stop is simply another pin driven through the plate. Sometimes it is the flattened bulb of a plate pillar, which, again, may have mounted on it a brass or steel plate to receive the hammer which has some springiness. Stops like these and other variations are not difficult to make or arrange, but their position is important since it limits the movement of the hammer tail and determines whether or not the hammer shank has to be bent to strike the bell satisfactorily.

Somewhat after the standard straight spring was developed, springs which acted also as stops were introduced, probably early in the eighteenth century. Typical shapes can be seen in Plates 8 and 25. Essentially, the vertical face of a normal straight spring acts on a tail to the hammer by means of a right-angled strip whose upper face supports the flattened boss of the hammer (Fig 39). It is not a dead stop, but it does not permit much rattling and bouncing, which is the great danger with sprung stops. The arrangement is excellent when properly adjusted, but it develops wear which leaves the hammer loose. The first essential is to bush the hammer arbor holes, after which the top of the spring can be faced with a piece of mainspring. These angled springs

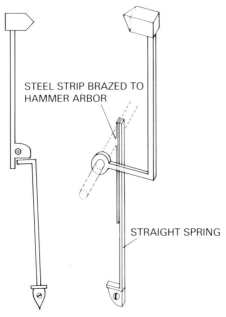

STEEL STRIP BRAZED TO
HAMMER ARBOR

STRAIGHT SPRING

a) STANDARD   b) VARIANT (SEE PLATE 14)

Fig 39   One-piece hammer springs and stops. An example of (b) is shown in Plate 29

149

are best made in two pieces. An unusual form of dual stop and spring is illustrated in Plate 29. The usual straight spring has been repaired, and behind it is a steel strip brazed to the hammer arbor so that the spring presses against its length. The strip projects only slightly above the arbor, so there is limited resistance to the hammer's being pulled back, but it projects well below the arbor so that the hammer has great difficulty in advancing beyond the position of rest, yet is not rigidly stopped. There is, of course, some wear at the points where spring and strip rub, but the action is very good even with the damaged spring.

In about the middle of the eighteenth century, bracket-clock cases began to have lower tops and soon bells were being mounted on the back plates. This meant that the hammer arbor had to be extended through the back plate and mounted on a separate cock, and often the stop arrangements were also taken out onto the back plate whilst the usual leaf-spring was used inside (Plate 27). The stop is then of a springy nature, allowing the hammer a good deal more movement than hitherto. The typical development of this type of stop is seen in Plates 27, 74 and 76, where even more movement is allowed – onto a gong – by the use of a long spring as stop for a pin on the hammer arbor.

French clocks used much lighter springs and stops. By far the commonest spring is a wire of blued steel attached to a screw which is pushed in through the front plate; the wire is so bent that it catches and presses upwards a small hook on the hammer arbor (Plate 16). Often there is a duplicate spring for the rackhook and locking piece (Plates 16, 33 and 34). Although these springs are and should be tapered, a straight wire silver-soldered to a screw is adequate as a replacement; the springs themselves are almost invariably tempered to a dark blue. In earlier clocks brass-wire springs bent back on themselves often

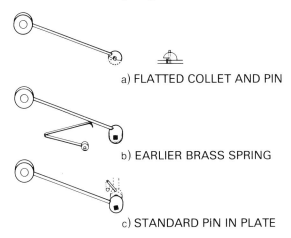

a) FLATTED COLLET AND PIN

b) EARLIER BRASS SPRING

c) STANDARD PIN IN PLATE

Fig 40   French hammer stops

150

provided the flexible stop. Later it became standard procedure to make a flat on the hammer boss which engaged with a pin in the plate (Fig 40). This seems never to have been general in carriage clocks where the alternative of a pin in the arbor meeting a pin through the back plate (Plate 35) was used. This was a rigid stop, but there was enough flexibility in the thin brass hammer shanks.

On German and American skeletonised early mass-produced clocks a coiled spring was wound round the hammer arbor and hooked round the edge of the plate or pushed through a hole to anchor it. The stop was by a wire extension from the arbor locking on a plate-pillar – economy, as always in these movements, prevailed. The spring for older cuckoo clocks was also a coil round the arbor. As the hammer arbor was the topmost of the three sounding arbors, the spring was merely

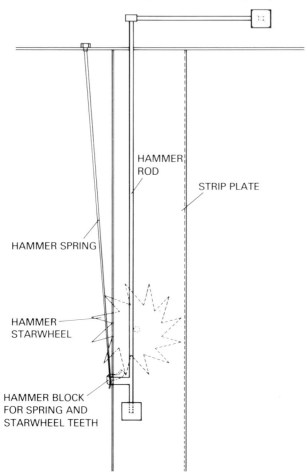

Fig 41   Traditional Comtoise hammer and spring

151

anchored round the arbor below at one end and round a wire at the other, the stop resting on a pin in the plate. None of these foreign systems presents much difficulty in repair.

Comtoises from about 1890 on may strike on gongs. They have a heavy hammer with a very powerful spring mounted on a sideways extension of the plate and after only a century or less, the hammer holes are found to be badly worn. The more traditional Comtoise, with overhead bell, employs a horizontally swinging hammer inside the bell, mounted on a long rod pivoted in cocks riveted to the strip back plate. The rod is turned by the pressure of a starwheel tooth on a nib fastened to the rod, with which a long tapered spring, descending from the top plate, engages (Fig 41). The spring is bent from a screw, like a normal French hammer spring on a gigantic scale. As already said, this spring is very brittle. Once broken, it cannot easily be repaired and the only option is to silver-solder very stiff spring wire to a screw and blue them as a replacement.

Chime hammers on older clocks are normally provided with individual slip springs working downwards from an upright frame (Plates 51–4 and 73–6), and shaped so as to bank on the back of the slots in the block where they are pivoted. This means that they have to be set very close to their bells as there is very little give in their short shanks. The springs may be riveted or screwed and, as they are individual, their replacement is no great problem. However, the springs

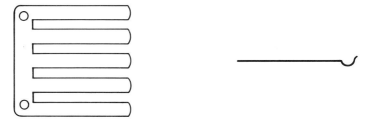

Fig 42   Modern chime hammer spring

of chiming clocks of the 1920s and 1930s and, indeed, now, are not individual but stamped out of a single strip of spring steel (Fig 42). Sometimes there is a set for the hammer blocks to check on as well, or else reliance is had on the flexibility of these much longer hammers. Occasionally there is even a third set fitting over the blocks to quell any tendency to bounce. It is not practicable to replace these strips of springs with individual tongues, and it is not easy to make a tidy job of replacing a whole strip. If one has a stock of spares, of course, one may well find a reasonable substitute. Otherwise it is a case of persevering with fine snips and shim steel.

152

Hoopwheels

The 'hoop' is a strip of brass some 5mm (⅕in) wide bent into a circle and fixed by means of lugs to its wheel (Fig 43). Sometimes it is missing, and sometimes it has worked so loose that there is a slot down which the locking piece can fall. In the latter case it can often be re-riveted. If one has to make a new hoop, this can be done from a strip of brass, which is rather difficult to hold, or from a slice of brass tube if you have one whose diameter fits the riveting holes in the wheels. Either way, thin strips must be cut out with a piercing-saw to leave lugs standing proud for riveting, and care must be taken that they line up exactly with the holes. If tube can be used, this sawing is much simplified since the tube can be held in the vice and the slot cut out only when the section has been removed. Otherwise, it is simplest to go round the wheel with a piece of paper or card marking positions for the holes, and then to cut out for the lugs before shaping the strip into a hoop. The slot should allow play for the locking piece arm as well as its head. Sometimes one finds that the leading edge of the hoop has broken down due to the sliding of the locking piece up it, and a pin has been placed in front. This seems to be a long-standing method of repair and perhaps one may use it if the hoop is otherwise good.

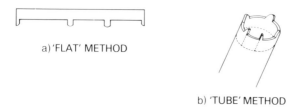

a) 'FLAT' METHOD

b) 'TUBE' METHOD

Fig 43   A hoop in the making

Lifting Pieces

These parts are often subjected to a great deal of abuse; one cannot insist too strongly that it is folly, and causes endless further trouble, to bend them except to straighten them when they are clearly distorted. If you bend the lifting piece, you effectively bend the warning piece, which is carefully adjusted relative to the locking to ensure that the train is not freed fully before the hour.

Many lifting pieces are unprotected against pushing the hands backwards. The hook type (Plates 4, 19 and 21) will eventually jam on a pin or star-tooth and then either be bent or break the pin; the typical French lifting piece (Plate 15) will be forced backwards out of adjustment with the warning piece. If these pieces have to be

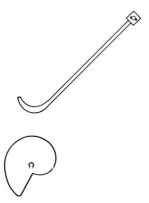

Fig 44   Unusual eighteenth-century 'safe' lifting piece and cam

straightened, the timing of lifting and warning has to be set up afresh by careful adjustment of the lifting piece – the lifting piece in Plate 5 has evidently been bent and readjusted more than once. The internal-rack movement in Plates 28–30 uses a lifting piece with reversed hook so that the surface presented to the lifting cam is curved and the lifting action will occur whatever the direction in which the hands are revolved (Fig 44). Far more typical, however, are the pieces shown in Plates 23, 26 and 73. Here the lifting piece is riveted to the warning piece and is of hard springy brass turned on its axis so that the lifting pin will push it harmlessly outwards if the hands are reversed. Detail of a typical modern protected lifting piece is shown in Plate 67, where the lifting and warning levers are separated from an angled lifting piece which is sprung and capable of being moved in either direction without damage. This is the system also in Plate 59.

In let-off by lifting piece, as opposed to flirt, warning takes place during the sliding of the pin or cam along the piece, and as the pin drops off the piece the train is finally released. Consequently the precise length of the lifting piece is not something to tamper with. If the striking or chiming do not correspond to the position of the hands, either the lifting pins, if any, must be bent or, preferably, the minute hand must be moved on its collet, if it has one. Obviously if all is well at the hour but not at the half, or at some quarters but not others, it must be the lifting pins rather than the lifting piece which are at fault.

Pinwheels

A missing pinwheel is straightforward to make. It is quite satisfactory to drive the pins in with a little Loctite 601, or indeed, dry, and the scale of the whole will tell you the size near enough. It is wise to make use of a hollow punch in sizing, and the pins should be chamfered off again

154

(see Chime Barrels above). The only problem is the number of pins in a missing wheel and the number of teeth.

When the clock was designed, the number of pins was set partly by convention and partly by the need for adequate clearance for the hammer tail between blows. However, once it was made, that number of pins was built into the gear train and could not be altered. The locking wheel must be capable of locking after a single blow or sequence, and so makes one revolution per blow. It follows that its geared ratio to the pinwheel pinion which drives it must be the same as the number of pins. When the whole pinwheel is missing, so is its pinion, so that you do not strictly know that ratio, but you can calculate it from the number of teeth on the locking wheel and the space left for the pinion, whose teeth must be the same size. In fact the pinion is virtually certain to be of 6 or 7 leaves and observation of other pinions in the movement will probably tell you which. You may also know that 8 pins are normal on eight-day clocks, and 13 on the greatwheel of a thirty-hour clock, so that, all in all, the problem of a missing pinwheel is not serious so far as the number of pins goes. The number of teeth on the missing wheel can be calculated from the space available, but equally from the requirements of a striking train set out in Chapter 2. The same considerations govern the number of teeth on a hammer-lifting starwheel but here, in clocks of the 1920s and 1930s, 10 teeth were general. If fitting a new pinwheel, adding a set-screw to simplify setting up is strongly recommended – not in an antique, however.

Racks and Rack Tails

The geometry of the rack system is crucial and is based on circles from the pivot points (studs). First, with regard to the rack and snail, there has obviously to be a fixed relationship between the working length of the rack – the space containing all twelve teeth which are gathered – and the distance covered by the rack tail from its point of rest to the lowest point on the snail – the twelve o'clock section. If you know one of these distances (whichever one), you can work out the other. Basically, the relationship between them is that between the circumference of the circle at the rack pivot passing along the pitch of the rack teeth, and the circumference of the circle with the same centre which is traversed by the rack tail (Fig 45). Having found the ratio of these two, you multiply the smaller distance by the ratio to discover the working length of the rack, or divide the latter by the ratio to discover the height of the snail. In practice there is of course some clearance between the rack tail and the highest point of the snail. The working length of the rack can then be divided to give you the size of the teeth; they are usually two degrees each of the rack circle.

155

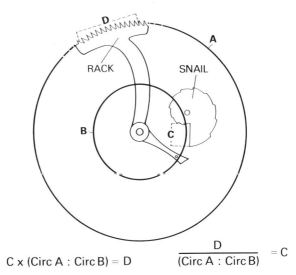

$$C \times (\text{Circ A} : \text{Circ B}) = D \qquad \frac{D}{(\text{Circ A} : \text{Circ B})} = C$$

Fig 45   Working length of rack related to working size of snail

The other vital dimension is the length of the rack tail, which is often found altered or is missing. This can be calculated from the maximum fall of the snail. If the working section of the rack and its centre are drawn as a triangle (Fig 46), the point where a horizontal line of the length derived from the snail's extreme height meets the sides of the triangle's sides indicates the radius of the rack tail's path, as measured from the stud to the tail-pin or nib. In practice the rack stud is often the centre of a circle on whose circumference lie the snail stud and the rack tail's pin, but this is a very rough and ready guide. If you have neither snail, rack nor rack tail, you must still have the rack stud or its hole,

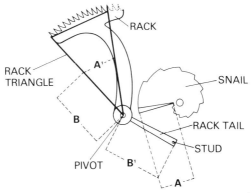

Fig 46   Calculating the length of the rack tail. Take measurement A, fit it to the rack triangle A' (working teeth only) and B (or B') is the length of the rack tail from pivot to stud

156

the gathering pallet or its arbor, and the snail's centre be it starwheel stud or central arbor, so that you can still reconstruct the basic geometry.

Whilst the working length of the rack can be calculated and divided, or the teeth can be measured in degrees, just as a countwheel can be drawn out in the abstract, there is much to be said for plotting the course of the gathering pallet with the rack set at each step of the snail, as a means of marking out the teeth. This can be done on a cardboard model of the blank rack – such racks, large enough for all normal purposes, being readily available from suppliers. The difficulty with division and marking by degrees is that errors become cumulative and the later teeth can end up some way out. This will, of course, be avoided if you have the facilities to clamp up the rack blank and cut the teeth with a fly-cutter made up to the desired width, when all your teeth will be the same automatically; but a perfectly adequate and decent-looking rack can be marked out by hand and cut by piercing-saw, with possibly a little adjustment by filing to the backs of the teeth. It is a very satisfying task, if somewhat slow. When cutting the rack, remember that at least the last tooth, and sometimes more than one, is not gathered by the gathering pallet but has to be held by the rackhook; therefore you must cut thirteen teeth, and fourteen or fifteen are quite usual. For an early British clock, the rack will be of steel; for a French or modern British clock it will probably be of brass. If the rackhook is present, its shape should be borne in mind when determining the precise shape of the teeth; it is more common for a hook to fail to hold than for a pallet to fail to gather, and some racks were made with considerably undercut teeth conforming to angled rackhooks.

The British rack will normally be riveted to a pipe riding on the stud, with the rack tail on top. This pipe must be a good fit for the stud, but need not necessarily be very thin – there are difficulties in mounting the rack tail, which has to be adjusted and then fixed, and pinning or screwing is preferable to riveting here, so that some thickness of metal is required at the top. The brass French rack poses similar problems but here the scale is usually too small to permit screwing on the tail. Incidentally, the arms of these racks, particularly in carriage clocks, are often very curved (see Plates 34, 35; contrast Plate 33). The same geometric principles apply, but the curve is ignored in calculating the length of the rack; it is partly aesthetic and conventional and partly to give clearance from the hour wheel. The angle in the rack tail in movements such as that in Plate 34 should also be ignored; it does not affect the circle traversed but only the adjustment of the rack tail's position.

Some rack tails are completely unprotected against failure of the clock to strike twelve o'clock, but the majority have some defence.

Basically, the whole tail may be flexible, or springy brass, or a steel-sprung section carrying the pin may be built in. In either case, the leading face of the pin is usually bevelled off, as may be the highest face of the snail (Fig 13). The inserted sprung pin is certainly preferable, since there is less chance of distortion and incorrect counting from the snail when it is used. On some modern clocks, eg the Enfield movement, Plates 65–7, the spring strip holding the pin extends right along the tail and is screwed in place, the pin working through a narrow slot in the brass tail below. This seems a very simple and effective arrangement to imitate where there is no strong historic or aesthetic argument against changing the system. If, however, an eighteenth-century piece is fitted with a simple tail of brass, the best should be made of what is there; updating what can, after all, work reasonably would not be justifiable. The ethics of 'restoration' have not been much discussed here because with striking mechanisms they are fairly clear cut; however, there are opportunities for vandalism in adapting modern parts just as there are for a degree of fraud in trying to recreate the original with intent to deceive.

The completed rack system must be very carefully tested. Even if the pallet hole has been bushed and the teeth are well cut, irregularities in gathering can occur from time to time even when everything is usually perfectly all right, and these inconsistencies have to be traced and removed. A great deal depends on having the correct rack tail length and on adjusting that piece correctly. The mechanism has to be taken through tooth by tooth, step by step, until that position of the rack tail is found when gathering (see Gathering Pallets above) and rackhook retention are smooth and correct. It will take some time to eliminate the rack tail as the cause of any trouble, and only then should attention turn to possible irregularities in the rack which adjustments to the geometry will not correct. In this case, you look for trouble recurrent on certain teeth and you may have to alter them. But remember that each tooth is acted on twice – by the pallet and by the rackhook – and it is no use correcting the pallet action in such a way that the tooth is not held by the rackhook. This is what happens if teeth are shortened or separated with a punch, so you have to proceed very carefully indeed, confining your attention as far as possible to the curved backs of the teeth and, above all, taking the affected teeth through both pallet and rackhook each time they are tested.

It sometimes happens that racks appear to be bent – and they quite possibly may be, especially rather delicate French racks with thin central stems. The symptom is that one half of the rack is gathered satisfactorily, but the other slips. The rack should be tested, as already mentioned, to see that the tips of its teeth do fall exactly on the circumference of a circle of which the stud is the centre. If they do not, adjustment by bending is probably called for, but remember first that if you bend one half of the rack in, the other half will come out – the proper adjustment is about half what it seems to be; and secondly that some tolerance is possible here depending on the stiffness of the rack spring and the adequacy of the rackhook. If a stiff spring and a rackhook which does not properly fit the teeth are satisfactory over one part of the rack and not another, it is they rather than the rack which need altering. The danger of altering rackhook profiles, especially the double rackhooks of quarter-striking carriage clocks, has already been mentioned, but the danger of modifying a rack is very much greater because metal removed cannot be replaced, and making a new rack is always a major undertaking.

The same considerations govern chime racks except that, being so much shorter, they are much less subject to distortion. The difficulties here are in adjusting the various springs so that they work in harmony. The strike warning-lever spring must not overrule the chime-rack spring, nor must the adjustable strike warning lever 'bottom' so that the rack cannot be fully gathered; the rack spring must be sufficient to knock out the hour rackhook at the fourth quarter, but not so strong that excessive strain is put on the gathering mechanism. These adjustments to bring about reliable action can take some time.

Rack Springs

Whilst one does occasionally come across twisted steel wire which appears to be original, the correct style of rack spring for most old British clocks is that seen in Plate 27. And Plate 23 shows very well how not to replace one that is missing. The rack spring has to exert even tension, as the rack moves a considerable distance, by pressing near the pivot point of the rack. For this purpose the bent-back brass wire is unequalled, and its brass foot, which should have a steady-pin to prevent it from turning, is both stylish and the best way to fix it. Such feet can be filed up and drilled quite quickly and the spring secured in the hole with Loctite. Springs are, however, available from suppliers

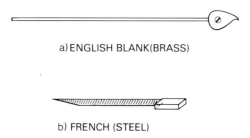

a) ENGLISH BLANK(BRASS)

b) FRENCH (STEEL)

Fig 47   British and French rack springs

159

ready-made and require only a little finishing. The bent up tip of the spring must be located where it can have some effect (Plate 27), not right up against the stud, but against the tail usually extended from the rack to receive it.

Carriage clocks, on the other hand, and many French clocks of the last 150 years, use steel springs with integral blocks secured by a screw and steady-pin and working on a pin descending from the rack (Fig 47). The spring is usually taken down into a point where it meets the pin. These springs are not available from suppliers and they are difficult to make in one piece. A reasonable repair is to solder a piece of watch mainspring to a steel block, filing and polishing until the joint is scarcely visible.

The force of the spring must take the rack tail down to the twelve o'clock section of the snail, but it need not do so violently – a common cause of maladjusted rack tails and unreliable striking. Neither should the spring be bent in such a way that it is stiffened to slow the speed of striking, for this can only in the long run lead to unnecessary strain and wear on the gathering mechanism. In all sounding systems a positive but smooth action without clatter and banging or jerking of the train, is what is required.

Snails

The geometry of rack systems (see Racks and Rack Tails above) will also give the maximum size of a snail. As with countwheels and racks, there are two ways of designing these – in the abstract or on the movement provided you have the rack. Again, it is preferable to do the job on the movement; there is a certain security in doing this which breeds confidence.

For the hour snail, you require a blank of its largest dimension, with a hole which fits closely over the hour wheel or starwheel pipe, either of which you may have to make first. Leave any screwholes, however, until after the snail is made. On this blank mark twelve radial lines, then push the rack up until it is gathered for one o'clock. Holding the rack, release the rackhook and let the rack fall one tooth and be held by the rackhook; hold it there, and use the rack-tail pin, sharpened if necessary, to make a mark across one of the radial lines. This will indicate the highest point of the snail. Do the same for the other operating rack teeth in order, and then extend your crossing marks on the blank snail until they are circles. You can then draw in the snail steps using part radial and part circular lines, cut out the piece with a piercing-saw, then tidy up with emery buffs. It is usual for the descending edge of each step to slope slightly rather than be radial, but this does not affect the working.

160

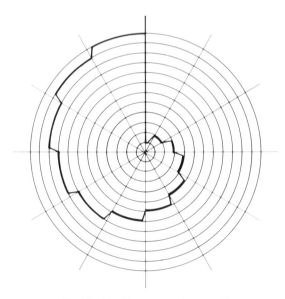

Fig 48   Marking out an hour snail

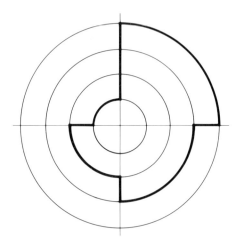

Fig 49   Marking out a quarter snail

Design in the abstract is no more difficult, in fact it is less awkward. But you cannot be quite so sure of a perfect match to the rack, which may not be quite regularly cut. Again divide your blank into twelve equally spaced circles, plus one extra for the arbor's ring and excluding the arbor hole; strike twelve equally spaced radial lines and then draw in the steps, cutting them with a piercing- or jig-saw (Fig 48).

The quarter snail, too, can be made from the rack tail, or abstractly in exactly the same way with four circles, except that the steps of the

161

quarter snail are left sharp and radial (Fig 49). Sometimes the lifting pins are fixed to the snail outside or inside – choose a blank of adequate thickness – and sometimes to the wheel, depending on the height of the rack tail from the plate.

Do not forget that with a ting-tang quarter-striking clock having one rack, which in practice excludes carriage clocks, provision has to be made for the rack tail to fall sufficiently for three quarters to sound after one o'clock. The first high step is divided into four equal sections, the last three being down at the level of the third step. It is best to make these snails on the movement. Make the hour snail first and then observe the fall of the rack tail required at the quarters and mark the snail at the lowest point – that for the third quarter. For safety's sake, there is something to be said for marking intermediate quarters as well; but this is seldom, in fact, done.

The star of a starwheel snail is made with twelve pointed teeth and is either riveted or screwed to the snail. It should be substantial, certainly not less than 1mm (30 mil) thick, as the teeth points wear. The position of these teeth in relation to the steps on the snail is crucial, since the jumper must hold the starwheel in such a position that the rack tail strikes early in each snail step. It is therefore best to make the snail a friction fit at first, set it in position with the sprung jumper and then adjust it by letting the rack fall, before fixing it and the starwheel. Equilateral teeth are satisfactory, but starwheel teeth are usually given curved backs and sloped away from the direction of rotation. This is to give the motion-wheel pin a vertical or undercut face against which to press when driving the snail, and so that the jumper slides relatively easily up a tooth and then swiftly tips over to flick it into position. The size of the starwheel is governed by the action of the driving pin, which should engage nearly half-way down a tooth. To assist this action, the jumper's angle should be trued up to a point; in course of time jumpers wear round.

Lifting snails or cams can be marked out in the same way as an hour snail, the steps being levelled out, if there is no half-hour point. If there is a half-hour, or quarter-hour, point the cam is based on a circle bisected or divided into four, with the divisions as the vertical drop-off faces. Where the hour lifter is longer, as in many chiming clocks, two circles are used – one for the quarter and the outer one for the hour (Fig 50).

Two minor features concerning starwheels should be mentioned here, although they pose no special repair problems and I have met them only on carriage clocks. The first is that the wheel may be turned by a 'finger' attached to the cannon pinion, rather than by a pin. The second is that the set-hands wheel is sometimes provided with a hoop into which a tooth of the starwheel can enter only at the hour. As a

162

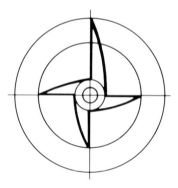

Fig 50   Self-correcting quarter lifting cam

result, the starwheel cannot be adjusted from hour to hour to suit the position of the hands.

Studs

The front-plate studs on which so many striking and chiming parts are pivoted should be polished and a good sliding fit for the parts – binding and free movement out of the vertical are alike detrimental. They are seldom interchangeable in length, thickness or even in screw-thread on older clocks; therefore it is always a wise precaution to set them out on a piece of card in their relative positions when dismantling the movement.

If you have a lathe, turn studs from silver steel, placing them in a vee-block to drill the pin hole, and turning a nick below the square since the die for the screwed part will not reach the square and the square must be seated properly on the plate (Fig 51). Older studs were

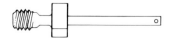

Fig 51   Turned stud, with nick above thread for clearance

often tapered, and clearly replacements must be tapered likewise or the fit will be sloppy. To ascertain the taper, force a piece of Plasticine, balsa wood or clay into the tapered part so that the moulded material assumes the correct shape, and then taper-turn or file accordingly. Where a very large tapered hole is involved and a lathe is not available use a straight stud, ensure that the tip is a close fit and use a brass bush, appropriately filed, to fill in the hole at the bottom of the post. A reasonable stud can be made in this way from straight stock, and tapped at the end to receive a square nut fixed with Loctite. If you have a lathe, however, you should use it, particularly for the threading, to ensure an upright piece. Similarly the hole for the stud may have to be enlarged if you cannot match the thread, and both this and the tapping are best done with a vertical drill to ensure that the stud is upright. Do not be satisfied with a poorly fitting screw action and do not use Loctite to fit the stud in the plate. This is bad craftsmanship and will be a nuisance to later workers.

8

SETTING UP AND ADJUSTING

Setting Up

Whatever the mechanism – rack or countwheel, strike or chime – there are three essentials in assembling the train and putting on the lifting piece:

1   All hammer tails must be off the pins or cams both when the train is locked and when it is at warning (Fig 52).
2   The warning pin must have a run, usually half a turn but some-times less, before it reaches the lifted warning piece (Fig 53).
3   Where there is warning, the warning piece must be in position before the train starts to run so that the warning pin meets the warning piece (Fig 5).

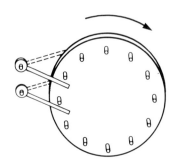

Fig 52   Hammer tails clear of pins

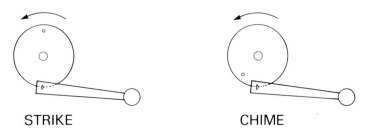

STRIKE   CHIME

Fig 53   The run to warning

165

In many French clocks separate cocks facilitate these adjustments after assembly (see Plate 17), and in modern chimers the use of set-screw mountings assists. But in any case it is always wise to set up the train in locked position, though in non-adjustable movements it is still often necessary to lift a plate later and correct the settings. Failure to follow the rules may lead to inability of the train to run, to unreliable counting and to striking when there should be only warning.

Hammer Tails
Where the hammer-stop is adjustable, ie it is not just the plate pillar, bending it permits small adjustment to the tail of the hammer after assembly; the same is true of bending the hammer. In some compact movements with small pinwheels it is not possible to allow both a reasonable run to warning (the distance from the warning pin's position of rest to where it meets the warning piece) and also freedom of the hammer tail from the pinwheel at warning and locking. In this case freedom when at rest comes first.

Rack movements pose less problems with hammer tails because there is some choice of position for the gathering pallet which governs locking, and one position for the pallet may well lead to a free hammer when others do not.

Setting up hammers and warning on an old internal countwheel movement can take some time; and inside chime barrels are particularly difficult because the hammers cannot usually be put in place until the movement is assembled, when you may have to start again.

Some modern manufacturers, including Joseph Kieninger (of Aldinghem, Germany) and makers of cuckoo clocks, use an accessible starwheel, rather than a pinwheel, with a set-screw fixing, which greatly simplifies this part of setting up. Chime hammers on outside, and some inside, barrels, can be adjusted later by means of set-screw ratio wheels. Early British posted thirty-hour movements and French Comtoises are particularly difficult, because the strong hammer springs and weight of the hammers are inclined to upset any adjustment just as it is made. I find it is simplest to tape these hammers in position and not to release them until the train is safely assembled between plates.

It is wise to try gathering pallets temporarily in locked position with the racks during assembly of the train, and any countwheel must be in the locked position with its detent – fastened there if necessary. Eighteenth-century chime countwheels are not adjustable, are mounted on squares and have to mesh accurately with their countwheel detents when the train is locked. As the detents cannot be fitted until the front plate is on, it is almost certain, with these clocks, that some juggling will have to be done by lifting a plate after assembly.

166

That being so, it is best to fit the detent as soon as the plates are together and not to assemble the rest of the front-plate mechanism until detent and countwheel have been tested, for the detent is obscured by other levers and cannot be removed separately later.

Inside countwheels are best set up as locked at the slot before twelve o'clock. You have to be very careful about the freedom of the hammers and positive locking, with the detent correctly in the countwheel slot; there is no chance to adjust later and it is all too easy to turn essential parts while locating their pivots in their holes.

It is wise not to take much notice of previous repairers' scratches, which may be contradictory. However, a guide to follow with French pinwheels and locking wheels is a clear punch mark between wheel teeth and the corner filed off a pinion leaf. Mesh at this point when setting up. (The same applies to motion wheels.)

Warning

Excessive run to warning will put the hammer tail on the pin. Inadequate run to warning may mean that warning fails to occur. A safe rule for striking trains is half a turn's warning, but for small French and carriage clocks, no more than a quarter may be possible. Similarly, chime trains have a much shorter run, about one-eighth of a turn, which in the typical modern clock amounts to about 1cm (²⁄₅in).

A common trouble with chiming clocks is the sounding of one hammer at warning, due to misplacement of the hammers when at rest or to too big a run to warning. Another complaint is erratic chiming or striking before the hour. This is due to the simple fact that the warning piece is not properly in place when the train is released. Adjustment of the link between lifting and locking pieces in countwheel movements, and of the closeness of the warning piece to the rackhook in rack movements, is the cure for this trouble, which may well have been brought about by well-meaning straightening out of the 'bent' pieces during cleaning.

With the wheels, hammer, probably hammer spring and other between-plates parts in position, and with the going train, too, positioned in the hope that all is well so far, put on the front plate, lifting piece and cannon pinion, the minute wheel if that operates the lifting piece, and do a trial run of the sounding trains. Outside chime barrels are not needed at this stage, but the outside countwheel and pinion are, as are rack(s), hook(s) and snail(s). Test up to warning action first on each train, using a screwdriver or strip of metal as a substitute for the hour warning in a chiming clock. If all is well, add the hour warning work and try the full sequence with locking. At this stage you are checking primarily that the hammer tails are not lifted early and that the trains are not released prematurely at warning.

167

See that, when the lifting piece is raised by turning the cannon pinion or minute wheel, the warning pin drops onto the warning piece with no hammer movement. If there is movement, lift the plate and readjust the pinwheel; several attempts may be needed. At warning in a countwheel mechanism the hoop detent or locking piece should be out of its slot or disengaged from its pin, and the countwheel detent clear over its wheel. Apply power by hand. If the warning pin slips past the warning piece, bend up the tip of the rackhook to increase delay between raising the warning piece and releasing the train, having first checked that the warning piece is lifted high enough to come into the path of the pin. Similarly, in the countwheel movement, lower the link piece between lifting and locking arbors. In the French system (Plate 15) this is effected by flattening the angle of the curved piece which rests alongside the warning piece. In the countwheel system, make sure that full locking is being achieved and that the bottoming of the countwheel detent is not preventing the proper falling of the locking piece; if it is, bend the detent upwards a little. If, although warning is satisfactory, when you take the lifting wheel round further it jams as the warning piece hits the top of its slot, the warning piece must be lowered at least sufficiently for the lifting piece to clear the lifting pin, and preferably rather lower – you would have to take it much lower to spoil the warning action.

There are two forms of adjustable warning piece, apart from the hour warning of a rack-chiming clock where the lever usually has a friction joint. On modern countwheel-chiming clocks the warning piece, if it is not a rigid backwards projection from the strike flirt, may be mounted on the lifting-piece arbor by a set-screw, in which case it is adjusted to just catch the warning pin when the lifting piece is at its highest point. On the classic rack systems, the warning piece can be bent, but may be friction-tight on the same pipe as the lifting piece, in which case it can be turned downwards slightly if it is jamming. Old French lifting pieces mounted on the left do not have a friction joint. If they should jam – a bent lifting pin is the most likely cause – they can be bent, but it should not be necessary. It is certainly preferable to adjust the angle of a warning piece to its lifting piece rather than to file it down, for it must once have been correct.

Warning pins must, of course, be firm. Warning pieces and lifting pieces both take a sliding motion and ought to be polished. The warning-piece blade should be so angled that the warning pin remains still as the piece rises under the lifting action. Bent in one direction, it is less able to hold the pin; bent in the other, it will be trying to drive the train backwards.

Countwheel Striking

In checking over countwheel striking, bear in mind the essential principle of the system, that the train is locked only when the slots on countwheel and hoopwheel (or locking-wheel pin) and their detents coincide. The gearing on thirty-hour countwheels is very often crude, and you should test the countwheel in various positions relative to its driving pinion until the strike locks at each hour. Adjust the tension of the spring clip on the external countwheel; it can cause failure from friction if too tight. Sometimes there is a peep-hole in the countwheel through which the pinion can be observed, and the correct engagement is when the punch-marked or filed pinion leaf is visible through this hole. In French clocks where the countwheel is mounted on a square, it is still worth trying out the possible positions. It is most unlikely that the countwheels in these clocks require alteration and you should firmly resist the temptation to tamper with slots with a file; the most likely consequence of altering one slot is to make others incorrect too. Make sure that the detent of a thirty-hour clock is not bent, or so long that it bottoms on the wheel and prevents proper locking. It should be positioned radially in the centre of an hour slot. The French detent is short blued steel and can hardly be bent. It should land at the beginning of a slot for the hour, so that another movement can be effected in the same slot for the half hour. Although the detent and locking piece are firmly fixed to the arbor in French movements, the curved link piece, which rests on the warning-piece arm, can easily be loose on its squared arbor and this will make locking capricious. The general rule is that where there is a slope to give lift on the locking piece, as in the thirty-hour longcase, there is no slope on the countwheel slots. Where there is no lift available in the locking (pin locking), there is lift either in the countwheel slots in French countwheels or in the detent in pin-countwheels. This lift is necessary to get the train under way, but it should never compromise the locking; the countwheel does not lock and must not be allowed to obstruct running – its job is simply to count.

Countwheel Chiming

The principles are identical to those of countwheel striking, with the complications of letting off the hour and, possibly, self-correcting arrangements. At the fourth quarter, either a pin or a specially raised countwheel section, arranges for the hour warning to be fully operative and for the hour rack to fall. Both are done by the action of the strike flirt on the countwheel, and should occur about half-way through the fourth quarter's chiming so that the rack is properly down and the warning primed at the hour. Occasionally, modern versions incorporate a gradual-release arrangement for the rack. This involves an extra tail sliding round the revolving countwheel's pin and poses no special problems. It is designed to avoid the 'clang' of the falling hour rack and also to save wear on the rack tail pin.

Should the hour not be correctly released, it is not normally possible on modern chiming clocks to bend the strike flirt, since this is integral with the strike warning and locking mechanism, and to bend the lever itself would upset these arrangements. It may be possible, however, to bend up just the projection which acts on the rackhook. Alternatively, one can stretch the raised countwheel section slightly with a hammer peen or, with the same effect, bend down slightly the pin on the flirt which acts as countwheel detent; but the latter is not possible unless the locking piece is adjustable on its arbor to compensate. Older internal countwheel chimers may let off the hour by means of a pallet on the front plate attached to the countwheel arbor; all that is needed here is careful positioning of the pallet so that it does not act too late in the fourth quarter (Plate 59). The hour may also (Plates 55 and 73) be let off by an extension of the lifting piece, which is raised further by an eccentric hour lifting pin such as later is associated with self-correcting devices. If adjustment is needed, which is unlikely, the hour lifting pin can be bent. Similarly, if a self-correcting device fails to release at the hour, the long lifting tooth needs stretching – a light tap at the back with a punch will suffice.

Aurally, the pause between the fourth quarter and hour striking is a sensitive matter. Once the strike warning has been released by the locking of the chime train (ie the fall of the countwheel detent and strike flirt to which the chime warning piece is attached), the determining factor is the run of the released train before a hammer is dropped. The setting of the rack release and warning, as just mentioned, have no bearing on when actual striking begins. The interval will have been set basically when setting up the hammer tail off the pinwheel or starwheel, but it may be adjustable within limits if you can move the gathering pallet on its arbor. Minor adjustment can also be made by varying the depth of engagement of the pin and warning piece in the strike train. If the striking is much too early, it is likely that it is being released at warning, and the clearance to the locking piece or rackhook requires adjustment as already outlined under Warning.

Rack Striking

The most common faults are jamming, non-stop running, and incorrect counting. Jamming, if the pallet and rack are sound, is usually caused by maladjustment of the rack tail to the snail. Non-stop running may be due to a loose or missing gathering pallet, to failure of the pallet to gather at one or more points on the rack or, and this is very common, to failure of the rackhook to hold, with the result that the rack keeps slipping back. All these faults may occur on chiming or striking trains. Miscounting may be due to these causes or to an obstruction to full fall of the rack.

Check that the rackhook tip matches in shape the gap between the rack teeth and acts as a one-way ratchet, that it moves freely on its stud, and that the pallet is not so worn that it is collecting unreliably, especially if the rack spring is stiff. The pallet should collect one tooth and a bit more, so that the hook falls back behind onto the rack.

Poor gathering action and unreliable counting especially, after a certain stage, round the snail, may also be due to an incorrectly adjusted rackhook or one of the wrong length. On the early British and some French types, the rack and rack tail are friction-tight at their angle and the rack tail can be knocked out of adjustment by frequent jolting onto the snail. Once the correct setting has been found, they should be riveted more firmly or, if there is enough metal, pinned or screwed together. The snail should be so placed that the rack tail lands in the middle of a 'step'. If it is so set for both twelve o'clock and one o'clock, all must be well unless the snail is poorly cut. The only satisfactory test is to take the strike right through, establishing that, for every step of the snail, the rack tail is properly placed, the rackhook falls cleanly into the gap between the rack teeth, and the pallet gathers adequately. If the hook does not occupy a gap cleanly, adjust rack and rack tail until it does, before fixing. All this applies equally to rack chiming and rack striking. With a modern cam and pin pallet, ensure that the pin is vertical and cannot catch the rack as it falls, and that the cam lets the rackhook fall into the rack teeth before the pin pallet has released its tooth. This is a matter of adjusting the pin.

Attention has also to be given to the locking angles of rack trains, either of the gathering pallet with its pin on the rack or, in the modern versions, of the straight face of the gathering cam to the bent piece on the combined rackhook (see Chapter 7: Gathering Pallets; Lifting Pieces).

Where the hour hand is not fixed to a snail, you can simply set up the rack tail to fall towards the middle of a snail section. Where the hand fits on a square, however, and the snail is fixed to the hour wheel, the setting should automatically be correct or at least satisfactory, since it is predetermined by the maker. It is advisable to test at one o'clock and twelve o'clock to make sure, especially if you have replaced a rack tail or tail pin. With ting-tang rack quarters, the rack tail lands exactly on the beginning of a snail section; this is necessary so that it can fall three

171

times on the lower sections after one o'clock, and adjustment must be correct or more than one blow will be sounded at one o'clock itself.

To summarise, it is not in the nature of racks to wear irregularly and irregular performance is most often explained by the rack tail. However, racks can be bent, both so that they no longer lie on the circumference of a circle from their pivot, and forwards or backwards so that different surfaces of pallet and rackhook are engaged. Bending offers an alternative to the rack tail as the reason for a rack gathering well up to seven o'clock or eight o'clock and then trouble setting in. Therefore it is as well to take off a troublesome rack, lay it on a really flat surface over white paper and compare it with a circle drawn from the stud hole.

Rack Chiming

The general problems of miscounting, slip and jamming are the same as for rack striking, but there is the added difficulty that the chime sets off the strike warning and releases its rack. The strike warning piece is moved into position, as in the countwheel system, at each quarter; but the train is released to warn only at the fourth quarter. This is done by a system of balanced springs which needs careful adjustment.

First, the quarter rack, when released for the fourth quarter, must fall far enough and hard enough to knock out the strike rackhook. The latter is of course adjustable, and must be set far enough back not to be knocked by the quarter rack at the third quarter, or striking will commence then. The rack spring which controls this action must not be so stiff as to impede a regular gathering action and smooth running of the quarter train.

Secondly, there is the mechanism by which the strike warning is set in position whenever the quarter rack is released. A pin, usually on the back of the rack, holds down the strike warning piece by pulling up against a sprung and centrally pivoted lever, the other end of which is the warning piece. When the rack is released, the lever falls and the warning piece see-saws up into position in the strike train. Nothing more can happen until the strike train is released by the knocking out of the rackhook at the fourth quarter, at which point the strike train runs into warning which is released for striking only when the quarter rack is again fully gathered. This warning lever's spring in effect works against the rack spring. It must be stiff enough positively to effect warning when released, but it must not be so strong that it impedes the rack's gathering. In rack quarter chiming, the setting of strike warning and the knocking out of the hour rack at the fourth quarter are therefore points to be carefully checked over.

The interval between chiming and striking at the hour is, again, governed by the depth of engagement of the warning piece and pin in the chime train, and by the run allowed to the striking train before its pinwheel drops a hammer. The depth of warning is usually adjustable at the pivoted warning lever, which has a friction joint. It may be possible to vary the setting of the hammer tail by moving the gathering pallet on its arbor, but more often one has to free the plates and reset hammer tail and pinwheel.

Chiming in General

The final stage of setting up chiming is to put on the barrel (or adjust it if, on an older piece, it is internal and not movable independently of the train), ratio wheels (if any) and hammers. This is, firstly, to ensure that hammers are down when the train is locked – particularly necessary with chiming trains where as a rule there is little spare power available and the train needs to be well under way before it starts raising a hammer; and secondly, to arrange for the train to be locked at the start of a chime sequence rather than in the middle.

Whether chiming is by countwheel or rack, the progress of the barrel is continuous like that of a countwheel; the chime can never repeat unless sequences are identical, as in ting-tang and scales. Therefore, provided that the chime starts to move only from the beginning of a sequence, it can always be run, or the barrel turned, until the right sequence for a particular quarter on the rack or countwheel setting is reached. However, particularly with an internal barrel, it makes sense to start setting up at the first quarter with rack snail or countwheel, and to fit the barrel correspondingly. This is not only to bring order into the proceedings, but because the first sequence is usually the most easily identified on the barrel where it and the last part of the third quarter – which is usually identical – are often marked by a dot or nick, or it can be recognised by a sloping row of pins since it is often in the form of a scale or downward progression. In older chimes, one may have to make several guesses before the starts of sequences can be identified. Unhappily, there is as yet no catalogue of tunes and chimes; indeed, their relationship to secular and religious songs, and the reasons for choosing them at any particular time, form a large area worthy of investigation. (See also Chapter 7: Chime Barrels.)

Having decided on the start, fix the barrel and ratio wheels, one of which will have been left loose if you have an adjustable modern clock; if not, you will have to lift a corner of the plate and juggle the barrel into the right position. It quite often happens that when the barrel is positioned so that the hammers are free at the start of the first quarter, where there may be a considerable gap on the barrel, they are not free between any of the other quarters. This cannot be tolerated, and you have to persevere to find a setting for the first sequence where the hammers will also be free between the other sequences.

You can now try the system under power. Run a full test of chime and strike and then adjust the hammers – particularly any chiming on bells – for desirable volume and freedom from jarring. This is normally a matter of slightly moving a nest of bells or of bending hammer shanks, and should not be neglected.

Self-correcting devices are either non-adjustable and present no problems of assembly, or are adjustable quite simply if you remember that their function is to impose secondary locking at the end of the third quarter's chiming. You must, however, be careful to place the countwheel so that its detent is in the slot before the fourth quarter when it is the extra-long lifting cam or eccentric pin which will next raise the lifting piece. The system can be tested quite simply by placing the minute hand at a quarter past, manually letting off the chime to run through the second and third quarters, then turning the hand through these quarters; no chiming will occur if the device is working. The clock should then chime and strike correctly at the fourth quarter as the chiming train will have been released by the long, lifting cam.

Motion Work and Hands

So far, we have simply put in the cannon pinion or minute wheel which activates the strike or chime, and used the snail to test operation of the rack. The problem now is to ensure precise matching of the strike or chime with the time shown on the dial, and procedure varies considerably according to the type of striking or chiming system.

Where half-hour or quarter-hour striking or chiming are concerned, there is nearly always a difference in the pins on the cannon pinion or minute wheel which causes let-off. As we have seen, the hour cam is longer in most modern chiming clocks to ensure that the self-correcting device is overridden at the hour by a specially high lift. Equally, a lifting pin may be placed further out for the hour, especially in French clocks, to ensure full release of the rack which is generally not allowed to fall with the smaller lift at the half hour. Such a pin or cam may also cause release of the hour rack in a chiming clock (see Plate 55). Such clocks almost always have a square on the cannon-pinion pipe so that, in one of four possible positions, the minute hand unless bent or loose will line up with the hour lifting pin.

Where striking or chiming are let off by the minute wheel rather than by the cannon pinion, however, the position of the hand has to be set by adjusting the wheel. Temporarily put the minute hand in place at twelve o'clock. Then turn the minute wheel, without turning the cannon pinion, until the strike or chime has just been let off. It is at this point – sometimes marked with punched dots – that the minute wheel and cannon pinion should mesh. Once so set up, they must never be moved independently of each other. It is a good idea, before going further, to put the dial temporarily in place, to ensure that the setting is correct and that the minute hand does point to twelve o'clock on the dial. The latter, after all, are not always exactly square with their movements, and it is a barbarism to resort to bending the hand later. Sometimes, where lifting is from the cannon pinion, the hand does not line up precisely as it should. In such cases the hands are almost always riveted to a brass collet with a squared hole and, by holding this collet on a square and turning the hand carefully, the setting can be corrected, after which the collet must be riveted more tightly. If the setting is correct at twelve o'clock, but not at a half or some quarters, the lifting pins have to be bent to correct the anomalies.

In many rack clocks striking 'one at the half', the minute wheel is provided with a pin to operate a right-angled lever which prevents the rack from falling at the half hour. This must be positioned relative to the cannon pinion in such a way that the pin raises the lever to its highest point at the half-hour, but in doing this remember that the minute wheel moves anti-clockwise. With quarter-repeating and grande sonnerie carriage clocks there is a similar lever, but in this case the lever must be arranged to release the hour rack, which is normally held up by it, before the hour. When sounding grande sonnerie, ie hours and quarters, this lever is manually set aside and the hour rack falls also at the quarters. In other ting-tang movements (see Plates 40–1), the corresponding lever either pumps the high-noted hammer back out of engagement with the pinwheel at the hour, or lifts its tail out of the way of the pins, so that the hour is struck only on the one, deeper, bell. A little juggling may be needed to make sure such levers have no effect at the first and third quarters, but no special accuracy is needed with the half and quarter striking carriage clock with repeat since the repeat lever (Plates 34–5) has a pin which pushes the obstructing lever out of the way of the rack.

When a snail is on a starwheel, it has to be moved by the cannon pinion's pin just before the rack falls. As a rule, this causes no problems; but with a squared minute pipe it means that there is only one position in which the minute hand can be placed on its square, namely at twelve o'clock as the starwheel turns. This in turn means that, where let-off is from the minute wheel, the minute wheel and quarter snail must be positioned so that the fourth quarter is in operation when the minute hand is at twelve o'clock. The shape of the jumper influences the point of rest of the starwheel, taking up its position between the wheel's teeth. Do not tamper with this save, if

175

necessary, to sharpen up the angle (see Chapter 7: Snails).

Finally, the hour hand must point exactly to the hour when the minute hand is at twelve o'clock. This is governed by its meshing with the minute-wheel pinion and the correct position may be indicated by punch marks. If the hour hand is fitted friction-tight by a split pipe, there is naturally no problem, for it can be turned to the appropriate hour when the dial has been fitted. Sometimes, although it is fitted to a square, the hour wheel is held by a friction washer to the pipe and again subsequent adjustment is then possible. If there is a starwheel snail, the correct hour can usually be selected by turning it – except for the brief time when it is engaged before the hour and in those few cases where a device is included to prevent turning save on the hour (see Chapter 7: Snails).

However, if the hour hand is on a squared pipe, even if the starwheel can be turned the hand will probably not be positioned correctly on the dial when it is tried. The angle is not easy to judge when putting on an hour wheel without a dial. The dial must then be removed and the hour wheel taken out of mesh with the minute pinion, then turned until it is properly positioned; at the same time, regard can be had to the position of any 'day of the month' indicator so that that is moved in the small hours. Several attempts may well be needed. If there is a squared hour pipe and the snail is fixed to the hour wheel, there is only one possible position for the hour hand, ie that dictated by the number of blows struck for the particular position of the snail. But a tooth or two's adjustment either way will nonetheless be possible according to where the rack tail is to strike the snail, and it should never be necessary to bend an hour hand to make it line up with an hour on the dial.

9

CORRECTING FAULTS

The causes of many of the commonest faults in striking and chiming have already been mentioned, but it may be helpful now to summarise recurrent problems and how to approach them. Troubles are broadly of four kinds: it won't start, it won't stop, it sounds early, it miscounts. These apply equally to striking and chiming of whatever system.

Striking or Chiming Fails to Start

Naturally you first ensure that there is power in the train. Then, by moving the minute hand or arbor, test that the let-off arrangements are in order. The lifting piece, especially of the kind shown in Plate 62, may be bent, or a lifting pin may be missing. The warning piece, often adjustable, may be so bent or adjusted that it jams on the slot in the front plate before the lifting piece is fully raised or released; in which case the going may stop as well. If the going has stopped between twelve o'clock and two o'clock it is very likely that the rack was released and not fully gathered at twelve; as a result its tail either jams against the upright face of the approaching snail or passes over it by a safety spring (Fig 13) which, if stiff, will nonetheless still stop the clock. Of course if, as in many modern clocks, the warning piece is adjustable and when down is not cleanly passed by the warning pin, there is no chance that the clock will chime or strike reliably. Similar difficulties are produced by bent warning and locking pins; all must be observed in the rest and run positions for a positive action.

In many French rack-striking movements (as in Fig 2 and Plate 34) the rackhook is displaced, to release the rack fully at the hour or partially at the half-hour, by a pin which engages with the warning piece as the latter rises. This pin is easily bent and may then not move the rackhook sufficiently to release the rack at the hour, with the result that the clock strikes 'one' at every hour. The rackhook pin is far more likely to be the cause of this trouble than a misplaced lifting pin in the cannon pinion, to which one's attention is naturally inclined to turn first for the explanation.

Where let-off is by some form of flirt mechanism (Fig 4), a balance is

177

required between the power needed by the going train to keep going and that needed to prime the mechanism so that it will free the rackhook when the flirt is released. The flirt spring must be sufficient to knock out the rackhook positively – and for a big chiming clock this means a powerful spring. If this then causes the clock to stop near hours or quarters, attention must be given to the train, and possibly mainspring, on the going side. In carriage clocks, where the flirt is a loose cranked arm whose lower piece should drop on top of the rackhook, the small spring which keeps this part bearing downwards may need attention if the flirt action is positive but the flirt fails to catch the pin; obstructions are also possible, and it may be that the pin needs bending upwards slightly. Lifting pieces and flirts must be perfectly free on their studs, or their arbors must be free. Ensure that strike/silent arrangements do not interfere with the lifting piece save when required. The old British 'pump' system, with its sloped block pressing on a sprung lifting-piece arbor is normally on a fail-silent basis; if for any reason the lever slips or is bent, the spring draws the lifting piece forward to 'silent'.

As we have seen, save in very old and rare pieces, chiming of whatever system generally lets off rack striking at the fourth quarter, and this may fail to happen. First establish that it *could* happen, ie that the hour train is powered and able to run, by letting off the hour strike manually. Then look to see how it is let off by the chiming. With rack chiming (Fig 27) let-off is started by the deep fall of the quarter rack at the fourth quarter, which knocks against an extension of the hour rackhook, whilst the striking is held to warning at every quarter by the disengagement of the pivoted warning lever from the pin on the end of the rack. Thus actual let-off occurs when the warning is released by the gathered quarter rack, so long as the hour train is free to run. The warning system is fairly reliable, but the release of the rack often requires adjustment by means of – within small limits – the fall of the quarter rack itself, the hour rackhook extension, and the strength or tension of the quarter-rack spring. The latter must not be such as to impede gathering, but should ensure rapid fall of the rack tail onto the snail at the fourth quarter. The hour rackhook must be pushed aside by this movement; sometimes it is riveted and in two adjustable parts so that the depth of engagement with the quarter rack can be varied, and these may have become loose; sometimes it has become bent. However, neither adjustment must allow the slightest chance that the hook will be released by the lesser fall of the quarter rack at the third quarter. Note also that, besides causing unreliable gathering action, a quarter rack which wobbles on its stud can cause puzzling inconsistencies.

With countwheel, as with rack, chiming, the actual let-off is caused by the return of the chiming system to rest. The quarter rack allows

178

the far end of the hour warning lever to fall; the countwheel's detent, linked to the strike flirt and warning piece, falls into a slot and frees the hour warning. In either case, if release occurs early, there is sure to be something bent awry, and possibly also the depth of engagement of the hour warning piece is insufficient. The rack hour warning lever may be in two pieces, like the extended rackhook, and these can work loose. The pin on the end of the rack engaging with the warning piece offers further scope for adjustment. Trouble with countwheel mechanism will often be found to be due to tampering or rough handling, by which the countwheel detent piece or pin is bent and then the strike flirt twisted in an effort to adjust the locking. Thus what is really amiss is not the strike let-off but the chime locking, and this will have to be looked at first.

This applies generally to the more modern chimer with external countwheel and release of the rackhook by a pin or an extra-high section of the countwheel, but there are other systems for hour let-off in older movements with an internal-chime countwheel. The difficulty here tends to be with the pivoted hour warning lever, similar to that on rack chimers, on the top of the front plate. This is released by a lever fixed to the chime locking-piece arbor or by an extension of the lifting piece (Figs 24–5) and, as with most chimers, operates at every quarter but has effect only at the fourth. Its adjustment is quite sensitive and the piece is in rather a vulnerable position, yet it tends to be overlooked since it affects only one quarter in four. The chime train must be halted by the falling of the detent and locking piece just as the hour warning lever releases its pin, and small adjustments to lever and pin may be needed to secure this. There is also a small amount of timing variation built into the movement during assembly, namely the run which the pinwheel has before it actually raises and drops a hammer tail. If this run is inadequate or non-existent, there is no way by which there can be a tolerable interval between chiming and striking.

Finally, it should be added that failure to sound can be caused by all the usual faults in trains, particularly worn pivots and dry or dead mainsprings. But it is always worth a special check to ensure that the hammers are off their pins when at rest and that the fly is reasonably balanced, particularly in chiming trains.

Striking or Chiming Fails to Stop

For the owner there are few more alarming, and for the recent repairer few more humiliating, situations than the one where the newly serviced machine threatens to strike until it disintegrates from exhaustion. Unhappily also, owners demented by the abominable sound sometimes resort to extreme remedies like putting a duster in the mechanism or more logically but possibly more dangerously, taking off the striking weight but leaving the clock going. Fortunately, the causes of failure to stop are not usually difficult to find, being malfunctions either of locking or of counting or of both.

This trouble is most common in rack striking and is often caused by failure of the gathering pallet to pick up the rack because the rack is out of line, because the pallet is loose on its arbor, because the pallet is worn short, or simply because the pallet has fallen off – sometimes they are pinned on or secured by a nut, but more often than not they are simply a press fit. The remedies are generally clear enough. If the rack is out of line with the pallet, it may not be bent, but may have worn loose on its stud. Similarly, if the pallet fails to gather, that may be because its pivot hole is grossly enlarged; both these are common faults. It is not really satisfactory to pinch or punch the rack pipe more tightly; a new stud should really be made or the trouble will recur. Again, it certainly will not do to tighten this crucial hole with a closing punch; it must be properly bushed although the whole movement will have to be dismantled for the job to be done. However, although it is preferable to make a new pallet to replace a worn one, this is time-consuming work not necessarily required for all classes of clock. A satisfactory and reasonably presentable repair can often be effected by facing the pallet with a larger piece of spring steel. This should be carefully tapered and smoothed off so as to blend with the original metal – very little addition is normally required to restore normal gathering which, as already said, is achieved when the tooth of the pallet raises each rack tooth a little beyond the point of rest a tooth ahead, so that the rackhook falls back into its own position. Cam and pin pallets do not seem to suffer from wear, though pins break; perhaps they are not yet old enough. They are, however, mainly driven onto round tapered arbors and often come loose. The pins are brittle and should not be bent without careful thought. In replacing them, which is best done with pivot steel, the size is not critical so long as the rack is free to fall. The other main (associated) area of trouble is in unreliable holding of the rack by the rackhook (see page 159).

Failure of the English pallet tail to lock is rare; but if it occurs the cause is almost always either a loose rack pin or an unnecessarily stiff rack spring rather than wear. Pin locking, on the other hand, gives a fair amount of trouble. This is not generally due to the locking piece or pin – though either, if bent, will cause mislocking – but to the fixing of the rackhook to the locking-piece arbor (Fig 2). If the hook is at all loose on its square, it may come into position to arrest the rack, whilst the locking piece still fails to come into the path of its pin. The most reliable repair, short of a new arbor, is to stretch the square with blows from a light hammer, although filling in the rackhook hole with a wedge of metal may serve in an emergency. Often the square is tapered either throughout or at its base, and tighter pinning, perhaps with a collet added, will cure the trouble.

If the countwheel system fails to lock, for example because its locking piece is set too high or the vertical face (Figs 32–3) is worn or bent, the effect of the countwheel is overridden and continuous striking results. On the other hand, if the countwheel is detached but the locking satisfactory, the clock will always strike one blow only. Of course it is often not the locking piece, which is fairly robust in a thirty-hour longcase clock, which is bent, but the countwheel detent. If this bottoms too early on the base of a countwheel slot, it will prevent proper locking. If an external-chime countwheel is loose on its arbor, completely irregular (but not necessarily continuous) sounding occurs, since the let-off of the hours is also affected. Note also that the French countwheel locking (Plate 15) is subject to the same drawback as pin locking with the French-type rack mechanism – the link piece on the front plate can work loose on its square so that it fails to operate the combined detent and locking piece consistently.

In the same way that continuous striking can result from failure of the gathering pallet to pick up the rack in that system, so in the countwheel mechanism an almost complete disorientation of the countwheel and the locking or hoop wheel will result in at least a prolonged bout of striking. In this case, unless there is reason to think that unoriginal parts have been used, it must never be supposed that there is something wrong with the profile of the countwheel. To start filing out slots here and there means certain and growing trouble, for you will soon find another slot to attend to, and so on until the clock is locking twice each hour. Some external countwheels are so riveted to their gearwheels that they will work properly in only one or two positions, and you must patiently try until the right position is reached, having satisfied yourself that the locking wheel is not moving on its arbor. It is similar, if the options are more limited, with French countwheels mounted on squared arbors. Old locking wheels do not in fact tend to come loose, but the set-screwed cams used for locking in some more modern chiming clocks do, and these must of course be lined up with a notch in the countwheel and then fixed firmly, so that locking piece and detent action coincide.

With countwheel systems, above all ensure that the movement locks as it was intended to – on its locking piece. Locking or hesitating on the countwheel will very likely cause damage and make the striking impossible to let off or, if it does let off, progress to continuous sounding where the locking piece presents itself to pin or hoop just as the detent has mounted the next section of the countwheel. Where, as in the pin-countwheel type (Plate 11), warning, locking and counting are all worked from the same single arbor, considerable patience is required to adjust the three pieces to each other so that the train does not lock twice on a countwheel pin or, alternatively, fail to lock reliably.

A common and superficially mysterious cause of continuous or spasmodically incorrect chiming in countwheel mechanisms is any lack of freedom in the countwheel detent. This may be distorted, or the arbor slightly rusted; or the often-employed set-screwed collet may be pressing up too closely against the plate. Whatever the cause, the detent fails to fall into a countwheel notch when it should and, as its fall should release the striking which is held at warning while the clock chimes, the clock also fails to strike the hours. Needless to say, a locking piece on this arbor can produce the same effect if it is screwed up to lock the train whilst the countwheel detent is not in fact lodged in a countwheel slot.

Sounding Early

This is a really intolerable fault which demands instant remedy, its cause being a wrongly set warning. If a clock sounds early, incompletely, in ninety-nine cases out of a hundred it sounds at warning, and it sounds at warning for the simple reason that the train is unlocked before the warning piece is in position to arrest the warning pin.

There are several reasons for this situation. One is that on many modern clocks, especially on the chiming side, the warning pieces are adjustable. This simplifies setting them up, but may also permit them to move later. Another is that either the warning pin or the warning piece, on going or sounding side, is bent or worn. This, however, might not matter at all; it depends how finely the original adjustments were made. If at one time the warning piece only just held the pin, clearly a little wear or a jolt may lessen its chances of doing so, whereas with a deeper initial engagement the problem might not have occurred. Another possibility is that the run to warning has been set very short – particularly if shorter than on previous assembly – so that the pin catches next time when the lifting piece is that little bit higher. Combined with a hammer which has, whether or not of necessity, been left 'on the rise' at warning, this can be the cause of that rather odd situation when from time to time, especially with chiming where the various hammer-tail clearances can differ quite widely, we hear the first blow and no more, three or four minutes before the due time.

All this is of course quite distinct from the clock which comes for repair because it sounds completely 'early' all the time. That is simply a matter of adjusting the hands and the motion work, as explained in the previous chapter. What we are concerned with is intermittent and incomplete sounding, and it is something which never happens on clocks with flirt release, since they have no warning system. The raising of the warning piece into the path of the pin and the release of the wheel so that the pin drops onto the warning piece are the first two steps in the longish programme of events which constitutes the sounding of the vast majority of striking and chiming clocks. Therefore I would just stress again that it is vital to get this piece of timing right, by straightening the pin, bending the piece, adjusting its set-screw, or whatever will produce the desired effect with the least alteration to the clock. It really is quite as important a rule as the more often recited ones of 'run to warning' and 'hammers down when at rest', and it is all too easily left to chance in a clock which, before cleaning, gave no trouble.

Special note should also be taken of the critical nature of the adjustment in pin-countwheel movements. Their action is different from the usual in that, when the warning piece and lifting piece fall away to give let-off, the locking piece also falls, being on the same arbor, but to below the level at which it locks. The levels of locking, warning and running must be closely observed and adjusted, and the warning piece must be pinned firmly to its arbor. There is very little tolerance of the positions of the various striking parts in these movements.

There is one other simple reason for the sounding of a single admonitory blow. This is that, even if the hammer tails were all down – off their pins – when the train was at rest, they were so barely so, and the run to warning in a particular movement may be so long that a hammer tail drops off the pinwheel during the small movement of the train at warning. This is unlikely in the larger movements, but it can occur in striking carriage clocks where the clearance is very small, and it is not at all uncommon in chiming systems, where it is easy for a hammer to be left just up at one quarter but for all to be well elsewhere.

Miscounting

Much of this has already been covered, since continuous and early striking are due to the same causes – generally mislocking and maladjusted warning – as simple irregularity. Aside from failure of proper gathering due to pallet damage, and from a broken or fatigued rack spring whose effect will be fairly obvious, the commonest cause of miscounting in a rack movement must lie in the rack tail; for nothing can go wrong with the snail itself. An incorrect circle described by the rack tail must necessarily result in incorrect movement of the rack itself and so to unreliable gathering. As the rack and rack tail on a traditional clock were in any case set up friction-tight for purposes of adjustment, and may or may not have been permanently riveted once the matter was settled, the repeated banging of the rack-tail pin on the snail can lead to gradual loss of adjustment, particularly if the rack spring is on the stiff side. There is only one way to check and correct this and that is to take the striking – or, of course, chiming – round the whole sequence of hours or quarters, making sure that when the rack-tail pin lands at the start of the relevant section of the snail, the gathering pallet is poised to gather the correct tooth and the rackhook falls right to the bottom between two teeth. With more modern clocks, where rack and rack tail are in one piece, this difficulty seldom arises; though sometimes they are so shaped that they can become bent, so that the straightness of the rack-tail pin should be checked.

It seldom, but just occasionally, happens that it is impossible to achieve a perfect positioning of pallet and rackhook for both first and last teeth (ie for twelve and one). In that case, the rack tail is not of absolutely correct size, but it should be possible to find a compromise setting, as must once have been done, without mutilating an original rack or rack tail. If they are your own provision, that is a different matter, and you must adjust the rack tail's effective length. Too short a rack tail has the same effect as too small a snail – you will find that you get two 'elevens' rather than 'a ten and an eleven' or more probably you will detect similar double-counting earlier on. With too long a rack tail there will be skipping; but here be careful that you are not in fact suffering from a slightly over-sized pallet which can grab two teeth. Once you have corrected trouble with the rack tail, it is not unreasonable to add a screw or pin to fix it permanently; however, there often is not enough metal in the pipe to do this and in any case on a valuable piece such an alteration is undesirable. The alternative is to rivet the tail firmly. If there is any doubt, mark its position, remove it and add a drop of Loctite before riveting – slipping rack tails tend to be persistent.

Quarter-rack systems are subject to the same problems, though to a lesser extent. The main consideration is that the rack fall far enough to knock out the hour rackhook at the hour, but not far enough to touch it at the third quarter. If this is correct, the other counting will probably be so as well. Bear in mind also the requirement of the ting-tang (Plates 40–1) with its cut-outs in the snail for the quarters in the first hour. The snail must be positioned so that the rack tail strikes it at the very beginning of each section in these clocks or there will be curious sounds at around midday and midnight. With modern racks and pin pallets, ensure that the rack – usually unsprung and light in weight – is free and can fall clear of the pin pallet, for often it passes very close. With quarter-striking and repeating carriage clocks, check that the lever preventing the rack from falling at the half hour or quarters is properly set up, and is set aside by any grande-sonnerie control lever. Make sure, incidentally, that you do not happen to reverse the hammer-

184

controlling hooks (Plates 36–9) which work on the lever lifted by the hour-rack pin; they are not usually reversible.

Strictly speaking, incorrect counting with a countwheel cannot occur, unless the wheel was not designed properly. But we all know that it does occur, and that the form it takes is invariably failure to lock at a slot. This will not, in practice, be regularly one slot – though it is worth a check – because the fault is unlikely to be in the wheel and you should not, as already said, start trying to tailor it. The trouble is either in the locking (see previous section) or in the place where the countwheel detent lands, and how it lands, relative to the position of the locking assembly. The detent should never meet any obstruction at a gap except the bottom, and this means that its head should move radially towards the centre of the countwheel. If it is necessary to ease the starting by giving the detent a little lift, that will be taken care of either by the shape of the hoopwheel locking piece, which has a sloped front for the purpose (Fig 32), or by a slope on the countwheel-exit face – as typically with French countwheels and modern chiming count-wheels where there is pin locking which is unable to provide any lift. Nothing will be helped, and something may well be hindered, by bending back the countwheel detent at an angle away from the radius.

It is best to arrange the position of the countwheel – during assembly, if it is a fixed internal type – so that the detent lands in the middle of an hour or half-hour slot; there is then no chance of just managing to lock twice within a slot (ie division), or of bouncing over one. This can involve several attempts with old external countwheels, because there is a good deal of play in their gearing. The French countwheel detent should land in the middle of the half-hour or of the hour space, not at the very beginning or end of either. Make particularly sure that the detent never lands on a slope leading out of a slot, for this will certainly lead to mislocking at some time. In many cases, of course, design leaves you limited option in placing the countwheel; for example if it is mounted on a square. If none satisfies, look for trouble in the locking.

The positioning of countwheels is equally important with chime countwheels. Depending on whether or not the locking system provides lift, they have square or sloped slots. The tendency in the present century has been to use a sloped countwheel (Plates 62–9), which in effect provides a shaped resting-place for the detent; you either have it correctly set up – often merely a matter of a set-screw – or it looks hopelessly wrong. In diagnosing trouble here, the crucial point is whether or not the hour let-off still occurs in the right place in the chimes, but wrong by the hands. If this appears to be so, the trouble is in the locking; if it does not, and the hour is wrong relative to sequences, look for a loose countwheel.

185

You may still have countwheel, motion work and hands correctly set up and find the sequences wrong; most probably you will have the right number of notes but the wrong ones. In this event, you have to adjust the chime barrel to what is 'said' by the countwheel, either by loosening a ratio wheel and turning the barrel without the train or, unhappily in an older clock, by loosening the plates and attempting – no easy task – to do the same. This can require several attempts. Even if you know the chime and are certain that you have it linked to the right quarter on the countwheel, because the spaces between sequences are not always equal all may be well for two quarters, but at the third quarter you may find yourself with a note short or at the fourth you may take on a note left over from the third. Fortunately, there is no question of hacking at a barrel in the way in which one has seen some countwheels mutilated; you must simply persevere until you find the exact position of the barrel, relative to the countwheel at rest and allowing for play in the gears, when the hammers are not raised because the train is either at rest or at warning. In finding your way round a barrel, remember that the first sequence is usually a descending scale or pattern, and that in general all sequences end with the lowest note of the chime.

LIST OF SUPPLIERS

The following selection of some of the many suppliers will provide a starting point if you are looking for replacement parts and blanks:

UK

Cousins, E., 335 Green Lane, Ilford, Essex, IG3 9TH

Greville, Charles, Airport House, Purley Way, Croydon, CR0 0XZ

Hadfield, G. K., Blackbrook Hill House, Tickow Lane, Shepshed, Leicestershire, LE12 9EY

Meadows & Passmore, Farningham Road, Jarus Brook, Crowborough, East Sussex, TN6 2JP

Richards of Burton, Woodhouse Clock Works, Swadlincote Road, Woodville, Burton-on-Trent, DE11 8DA

Rose, R. E., 731 Sidcup Road, Eltham, London SE9 3SA

Southern Watch and Clock Supplies, 48/56 High Street, Orpington, Kent, BR6 0JH

Walsh & Sons, 243 Beckenham Road, Beckenham, Kent BR3 4TS

The following stock old movements and parts:

Allcroft, A., Bulshaw Farm, Little Hayfield, via Stockport, Cheshire

Biddle & Mumford (Gears), 36 Clerkenwell Road, London EC1M 5PQ

Olivers, 15 Cross Street, Hove, East Sussex, BN3 1AJ

Bells, single and in sets, are available from:

Whitechapel Bell Foundry, 32/4 Whitechapel Road, London E1 1EW

USA

The Cas-Ker Company, PO Box 2347, 128 East 6th Street, Cincinnati, OH 45201

Esslinger & Co, 1165 Medallion Drive, St Paul, MN 55120

The Gould Co, 13750 Neutron Rd, Dallas, TX 75234

S. LaRose, Inc, 234 Commerce Place, Greensboro, NC 27420

Marshall-Swartchild Co, 2040 North Milwaukee Ave, Chicago IL 60647

Mason and Sullivan Co, 39 Blossom Ave, Osterville, MA 02655

Swigart E & J Co, 34 West Sixth St, Cincinnati, OH 45202

Tani Engineering, 6226 Waterloo, Box 338, Atwater, OH 44201

FURTHER READING

UK

Allix, C. and Bonnert, P. *Carriage Clocks* (Antique Collectors' Club, 1974)

Britten, F. J. *Watch and Clockmakers' Handbook* (sixteenth edition, Eyre Methuen, 1978)

Britten, F. W. *Horological Hints and Helps* (fourth edition, 1929; reprinted Baron, 1977)

De Carle, D. *Practical Clock Repairing* (NAG Press Ltd, Eric Bruton Associates, second edition, 1968)

—— *Clock and Watch Repairing* (Hale, 1981 edition)

—— *Watch and Clock Encyclopedia* (NAG Press Ltd, Eric Bruton Associates, 1983 edition) (USA: Van Nostrand Rheinhold, 1982)

Gazeley, W. J. *Watch and Clockmaking and Repairing* (Butterworth, second edition, 1958)

Goodrich, W. L. *The Modern Clock* (North American Watch Company, 1950)

Grimthorpe, Lord E. B. *Treatise on Clocks, Watches and Bells* (1874, reprinted EP Publishing, 1974)

Harvey, L. and Allix, C. *Hobson's Choice* (Malcolm Gardner, 1982)

Jendritzki, H. and Matthey, P. *Repairing Antique Pendulum Clocks* (Edition Scriptar, Lausanne, 1973)

Penman, L. *Clock Design and Construction* (Argus, 1984)

Rees, A. *Clocks, Watches and Chronometers* (extracts from Rees's *Cyclopaedia*, 1820; David & Charles Reprints, 1970)

Reeve, C. B. *Making a Chiming Grandfather Clock* (Argus, 1980)

Roberts, D. *The Bracket Clock* (David & Charles, 1982)

Robinson, Tom. *The Longcase Clock* (Antique Collectors' Club, 1981)

Robinson, T. R. *Modern Clocks* (NAG Press Ltd, second edition, 1942)

Smith, E. *Clocks, Their Working and Maintenance* (David & Charles, 1977)

—— *Clocks and Clock Repairing* (Lutterworth Press, 1979)

Timmins, A. *Making an 8-Day Longcase Clock* (Tee Publishing, 1983)

Vernon, J. *The Grandfather Clock Maintenance Manual* (David & Charles, 1983)

Wilding, J. *How to Repair Antique Clocks* (Brant Wright Associates, 3 vols)

USA

Harris, H. R. *Nineteenth Century American Clocks* (Emerson Books, 1981)

Questions and Answers of and for the Clockmaking Profession (AWI Press)

Monk, Sean C. *Essence of Clock Repair* (AWI Press)

Palmer, B. *Treasury of American Clocks* (Macmillan Publishing, 1967)

Rudolph, J. S. *Build Your Own Working Clock* (Harper and Row, 1983)

Tyler, E. J. *American Clocks for the Collector* (E. P. Dutton, 1981)

Whiten, A. J. *Repairing Old Clocks and Watches* (Van Nostrand Rheinhold, 1981)

⌐ Other books and an up-to-date Buyer's Guide available from:
American Watchmaker's Institute, PO Box 11011, Cincinnati, Ohio 45211. Readers may contact them for a current catalog.

INDEX

190

repairing, 141–2, *141*

French clocks, 52–5, 57, 63, 67, 73–6, *53*, *74–5*; gear ratios 36, 39–40; *see also* Carriage clocks, Comtoise clocks

Gathering pallet: cam and pin type, 22–3, 88–9, 91; double, 36; gearing, 36, 37; internal rack, 67; making and repairing, 142–4, *143*

Gear trains, 11–12, 33–40; chiming, 36–8; striking, 34–6

Gongs: repairing, replacing, tuning, 129, 144–6; rod, 39, 85; speed of sounding, 38; tape, 52, 77, 96, 122

Grande sonnerie, 12, 29, 36, 79–83, 175, *80–1*

Grimthorpe (Edward Beckett), Lord, 13, 22, 26–9, 86, 94

Hadween, Isaac, 47–8, *48*

Half-hour striking, 25–7, *26*

Hammers, 12–13; adjusting, making, repairing, 146–7, setting up, 165

Hammer springs and stops, 41, 45, 47, 69, *43*; repairing, 148–52, *148–52*

Hammer tails, setting up, 165–7, *165*

Hindley, Henry, 69

Hobson's Choice, 27

Hoopwheel, 21–2, 34–6, 39, 44, *42*; repairing, 153, *153*

Hour let-off *see* Let-off

Junghans, Arthur, 97, 100, 102–5, *101*, *103*

Lantern clocks, 11, 41, 93

Lenzkirch (Germany), 87, 97

Let-off, 16–20, 23–5, 169–74, *23*, *24*; *see also* Lifting piece, Motion wheels

Lifting piece, 12, 16–19, 24ff; adjustment and repair, 153–4, *154*; Comtoise, 70–1; modern, 88; rack chiming 115ff

Link from lifting to locking, 43, 55, 57, 67, *42*, *46*, *55*

Locking cam *see* Hoopwheel

Locking (-piece, -wheel), 20–3; cam and pin pallet, 88–91; Comtoise, 75–6; internal rack, 67; on warning wheel, 57, 121; repairing, 138–40, *138–40*; standard rack, 61–3; *see also* Hoopwheel

Locking plate *see* Countwheel

Mainspring, 105, 178

Miscounting, 183–6

Month movements, 35–6, 45, *46*

Moonwork, 94

Motion wheels (-work), 12, 16; position of, 17; setting up, 174–6; *see also* Lifting piece

Musical clocks, 11, 12, 105

Passing strike, 11

Petite sonnerie, 29

Pinwheel: gearing 34–6, 154–5

Position of sounding train, 1, 11, 41, 47, 61

'Pull repeat', 27, 61, 93

'Pumping': barrels, 31, 85, 107, 122, 126; hammers, 81, 83, 175; lifting piece, 31, 178

Quarter chiming, 11–12, 23–4, 83, 93–112, 115, *23*; in Comtoise clocks, 73

Quarter striking, 11–12, 13, 25ff, 79–87, *82*, *86–7*; *see also* Carriage clocks, Ting-tang

Rack: principle of, 15–16, *15*; correcting faults in, 177–8, 180–1 183–5; position of, 61; setting up and adjusting, 165–9, 170–6

Rack examples (chiming): adjusting, 165, 172–4; Continental 120, 122, 125, *119*; early flirt release, 118, 120–2, *119*, *120*; Edwardian, 125, 125–6, *123–4*; nineteenth-century, 122, 125, *121*; standard warning, 115, 117–18, *116*, *118*

Rack examples (striking): adjusting, 65–6, 165, 170–2; carriage clocks, 73, 77–83, *77*, *80–1*, Comtoise, 69–73, *71–2*; cuckoo clocks, 91–2, *90*; internal, 67–9, *68*, *70*; modern, 87–9, 91–2, *88–90*; quarter-striking, 79–87, *82*, *84*, *86–7*; standard French, 73–6, *74–6*; standard long-case and bracket, 61–7, *61–2*, *64*, *66*

Rackhook, 16; double, 69; hour linked to chiming *see* Let-off; repairing, 78–9

Rack spring, 159–60, *159*

Rack tail, 16, 63, 65; adjusting, 171–2; designing and making, 155–8, *156*; double, 63

191